THE GOD PARTICLE

The Discovery and Modeling of the Ultimate Prime Particle

Ted Jaeckel

Universal Publishers
Boca Raton, Florida

The God Particle:
The Discovery and Modeling of the Ultimate Prime Particle

Copyright © 2007 Ted Jaeckel
All rights reserved.

Universal Publishers
Boca Raton, Florida • USA
2007

ISBN: 1-58112- 959-9
13-ISBN:978-1-58112-959-5

www.universal-publishers.com

Ted Jaeckel

The God Particle

The discovery and modeling of the ultimate prime particle and how it covertly underlies and is responsible for the properties of matter and the forces of nature

To Pat Peter and Mike

Coming up with new questions, discovering new possibilities, approaching old problems from new perspectives, requires a creative imagination and is the sign of new progress in science.
- Albert Einstein

I consider it quite possible that physics might not be based on the field concept, i.e., on continuous structures. In that case, nothing remains of my entire castle in the air, gravitation theory included, [and of] the rest of modern physics. – Albert Einstein

CONTENTS

PREFACE 13

INTRODUCTION 17

1. WHAT WE KNOW 19
   Insight into how little we really know.

2. WHAT WE THINK WE KNOW 21
   New knowledge shows no sign of convergence to simplicity.

3. THE WHAT HOW AND WHY OF PHYSICS. 27
   The categories of knowledge and what physics tells us.

4. WHAT WE DON'T KNOW 29
   Listing the fundamentals that we still do not understand.

5. THE HIDDEN UNIVERSE 39
   Introducing Zeron Theory and the covert Universe that underlies the one we see.

6. CLASSICAL NEWTON 43
   Why we use Newtonian Physics.

7. QUANTUM CHAOS 45
   How Quantum Physics has led physics into the quasi-paranormal.

8. THE PRIME PARTICLE 51
   Introduction to the prime particle called the Zeron.

9. THE COSMA AND THE AETHER 55
   Modeling the Cosma - a different sort of Aether.

10. ANY OBJECTIONS? 61
   Examining possible objections to the models of the Zeron and the Cosma.

11. MATTER MYSTERIES 65
   Modeling the atom.

12. WHEN IS AN ELECTRON? 79
   Modeling the electron shell.

13. PHYSICS PUZZLES 89
   Examining Young's double slit experiment and its derivatives.

14. THROWING LIGHT ON PHYSICS 97
   The underlying basis of Newton's laws. What causes Inertia.

15. THE PUZZLE OF THE SLIT MACHINE 109
   The answer to the single photon double slit experiment. Indeterminacy experiments analysed in light of Zeron Theory.

16. A NEW RELATIVITY 115
   A critical examination of Relativity. Establishing time as an invariable.

17. RELATIVELY QUANTUM 127
   A re-work of Einstein's Relativity equations to take the direction of observation into account. The establishment of the Cosma as the universal frame of reference.

18. REVELATIONS 143
   Why light travels at velocity c. Derivation of $E=mc^2$. Mechanism of increase of mass with velocity

19. DESCRIBING A PLANCK 149
   Discovering the meaning of Planck's Constant and deriving the mass of a Zeron.

Contents

20. THE TRUTH WILL OUT — 153
The mechanics of electricity flow, Beta particle production and Gamma Ray radiation and the meaning of Charge.

21. CIRCULAR ARGUMENTS — 163
Modeling atomic spin and magnetism. Why mass can be converted to energy.

22. COSMATOMICS — 171
Modeling Gravity.

23. THE COSMA'S BIG BANG — 177
The reason why the expansion of the Universe is accelerating, why some materials are radioactive and why Black Holes should be observable.

24. HOMEOPATHIC MEMORY — 183
A critical examination of homeopathic solutions and experiments.

25. IT HAPPENS EVERY DAY — 189
A look at some everyday phenomena in the light of Zeron Theory.

26. IS THE COSMA REAL? THE CLINCHER. — 195
Gamma radiation from space could be the evidence for the existence of the Cosma.

27. THE END IN SIGHT — 203
The universe in all its complexity and diversity may be interpreted in terms of one particle, one force, and one law. Einstein's castle in the air has collapsed!

Preface

This book hatched for more than 20 years. It began when my dear wife unwittingly bought me an American book on basic physics and then had to endure the ramblings of an absent minded engineer as he got to grips with all the problems that the book presented. The book was beautifully written in easy to understand language and it contained many fine illustrations to elucidate the content, but somehow every solution produced more questions than answers.

It wasn't long before I realised that many if not most theories were based on vaguely defined "fields" and usually ended in empirical equations or formulae. Why? Surely there must be more to physics? Was there no common basis for the many theories? It seemed that at the atomic level there was simply no explanation of even the simplest of problems, like what causes the electron to circumnavigate the atom, either in an orbital fashion or, even stranger, in a random sort of fashion as demanded by quantum theory.

Then came Relativity and the soul rebelled. As an engineer I live in a world that Einstein himself would have described as being "objectively real". If I push something it moves and if I keep on pushing it accelerates. To my mind Relativity was so counterintuitive that I had to believe that here was a fundamentally flawed theory. Worse was to come. In terms of sheer unbelievability, Uncertainty Theory and the outcomes of thought experiments based on it really took the cake! Parallel universes live/dead cats and the apparent ease with which the scientific community accepted this as gospel almost made me angry. Surely the Creator of the Universe would not have created such a bizarre universe. So now I have shown at least part of my colours as a biblical fundamentalist. I am in good company. Einstein for all his clever theories remained steadfast opposed to Uncertainty Theory stating that God would not choose to play dice with his creation.

I eventually decided that I would have to start from some fundamental and try to unravel what for me was a highly unsatisfactory situation. What would the fundamental be? There were really only two alternatives, a force or a particle. Because a particle could produce a force I homed in on the particle.

The God Particle

This was named the Zeron, being the smallest thing this side of zero. Perhaps if I found such a particle I might equally dub it the "God Particle" the elusive sub-nuclear particle that scientists around the world seek so avidly.

Thereafter came many hundreds of models of matter and the forces of nature built up from this imaginary particle. It was a process that was periodically interrupted by the need to make a living from a small business, but it kept on coming back to haunt me time and again. Ten years down the track and I hadn't made much progress. At about this time I discovered the Internet. Instead of being continually frustrated at the lack of information in the local library, I was suddenly confronted by a seemingly endless store of information.

Developing Zeron theory at once became much easier and faster. However the theory remained in the realm of the speculative until I came to have a good look at Planck's Constant. When I discovered the link between this constant and the Zeron I knew with certainty that I was on the right track. Here at last was a direct link to conventional physics, one that could be quantified and which at the same time solved one of physics most enduring mysteries. Other discoveries came fast and furiously.

It soon became clear that the atom became distorted as it was accelerated through the new Aether which I had named the Cosma. By taking Relativity down to this Quantum level it also became clear why Relativity and Quantum Theory could never agree. By inserting a term into the Relativity equations to take account of the direction of observation, that problem disappeared into thin air. I suddenly found myself in a new universe that made utterly good engineering sense. It was a very exciting experience. The rest of the book developed relatively easily . Model after model was constructed all bearing the hallmark of lucidity and objective reality.

At the end of the day what have we landed up with? Is this the ultimate theory of everything? Well, yes and no. Yes because we seem indeed to have discovered the God Particle. Yes because the theory operates at the most fundamental level imaginable. Yes because it solves so many of the enduring mysteries of physics. Yes because it provides an overarching theory of great simplicity always the hallmark of good theory. No because the theory is really in its infancy. Most of the models are based on the simple hydrogen atom and although the same principles apply to any atom, the devil lies in the detail. It is likely that complex atoms will make for models of mind-blowing complexity.

I have no doubt at all that this theory will be utterly rejected by the conventional scientific community. The fact is that a huge multibillion dollar industry has grown up around particle accelerators such as CERN and Fermilab. Even more billions are due to be spent on an international atom-smasher within the next decade. They seek this so-called God Particle, the all pervading but totally covert base particle of the universe. Should we have discovered this particle all would be lost and this industry would in all likelyhood collapse.

My hope for the future is not to simply publish a book and introduce a new theory, but to plant the seed of a God Particle Movement wherein people who become interested also become contributors to Zeron theory. To this end, readers are encouraged to do an internet search for Zeron Wiki. There is much to be done to develop this very elementary theory into a comprehensive Theory of Everything.

Introduction

This is not a highly technical book. Besides its more serious application, which is discovering a new form of physics, the book is also meant to entertain. It can be read start to finish without detailed reference to complicated formulae. While the introduction of some formulae into the text was unavoidable, the text explains their meaning and context so that they may be skipped at first reading without detracting substantially from the thrust of the book.

In this respect the book is unique, dealing with highly technical subject matter in, as far as is possible, a common sense and non-technical manner. This has been achieved by constructing easily visualisable models of matter and the forces of nature. It is also unique in its approach to the work of such great names in physics as Einstein, Michaelson & Morley, Young, Heisenberg, Bohr, Aspect and others. The book is pretty irreverent and pokes holes in theory wherever holes need to be poked regardless of the reputations of the scientist involved. This was done not through arrogance, but through the unassailable logic that the new Zeron Theory permitted. It opened many doors previously thought to be impregnable.

At the start of each chapter is a summary of what that chapter contains. This was a deliberate strategy to prepare the reader for what is frequently a revolutionary outcome within the chapter. It's a sort of mental preparation for what follows. Sometimes the outcomes surprised even the writer but this is what made the writing of this book so exciting.

All in all, the theory presented here covers an extremely broad range of physics from the sub-atomic to the cosmic. However the theory is really in its infancy. Notwithstanding, it does lay the groundwork for a fresh new look at all the theories of physics.

Chapter 1

What we know

It is a common perception that we seem to know just about everything about everything. In physics this is far from the truth especially as far as fundamentals are concerned. There's something very strange going on.

We know an awful lot. Mainly we know how to make things and make things work. Consider the modern world we live in. We know how to make microscopes that are so powerful that they can detect the atomic structure of a metal surface. At the other end of the scale, we know how to build telescopes capable of detecting a one-candle power light on the surface of the moon or detecting light from galaxies at the very edge of our universe. We can make spacecraft that enable man to walk on the moon or build a space station of gigantic proportions rushing around the earth in everlasting orbit. We know how to build ordinary cars whose inbuilt computer power exceeds that of the moon landing craft by a factor of 10. We know how to build supercomputers that can do 10's of millions of calculations per second. More frighteningly we know how to destroy the planet by manufacturing atomic bombs in such profusion that if detonated would signal the end of life on earth. Modern life is crammed with computers television sets, cell phones and all manner of electronic gadgets. The list is endless and the future is pregnant with soon-to-be marvels of man's ingenuity. Yes we certainly know how to make things work. Virtually all of these marvels have their roots in physics.

We seem to know everything about everything and our knowledge base grows exponentially. Surely there is little to find out about. And yet, is this really true? We are about to embark on a journey that will convince you that lurking underneath all we know is another universe, invisible, undetectable yet utterly real. It controls everything around us. It is responsible for every force of nature and every particle of matter. If we find it, the discovery will have the potential of revising all the current knowledge of the physical world we live in.

Chapter 2

What We Think We Know

We examine the current state of physics and the potential for expanding our knowledge as the availability of information explodes on the Internet. Mankind is evolving at an exponential rate. It's the evolution of the mind. But Physics is a moving target and information is soon outmoded. As new discoveries are made, there seems to be no sign of convergence of the new data towards a simpler over-arching theory. The more that is discovered, the more complex the theories become. Even the theory of the expanding universe, seemingly well established many years ago, was recently radically altered. In order to make real progress we will have to be willing to critically re-examine even the most accepted theories and make a fresh start to try and discover the mysterious sub-universe.

If we are to believe the anthropologists, mankind has been evolving to its present state over a period of 3.5 million years. The earliest examples of hominids bear little resemblance to modern man. The cranium was much smaller and the size of the brain was 400 cc. Modern man's brain (Homo Sapiens) occupies a volume of about 1400 cc. Of course that doesn't necessarily mean that our thinking ability has increased proportionately because scientists tell us that we only use 10 % of our 1400 cc and nobody knows what proportion the earliest hominids used. What is clear is that the brain seems to have abundant spare capacity. Numerous studies of people who have lost a complete hemisphere of the brain in an accident are testimony to the ability of the brain to take on extra work. The other half of the brain learns to cope with all the bodily and cognitive functions.

Evidence for the evolution of mankind from the primitive to the modern seems to be well established from skeletal remains. Analysis of stone-age tools shows a steady increase of knowledge and skill. Now however many millennia later, a completely different sort of evolution of the human species has taken place. It has taken place almost unnoticed. It will not be found in skeletal remains or in measuring the size of the brain. It is to be found in the evolution of the mind.

Before the invention of writing or hieroglyphics, the repository of all knowledge was in the memory of the individual. Knowledge was basically acquired and remembered by the direct experience. A certain store of knowledge was also passed on verbally from parent to child. Much of this knowledge got lost along the way or was forgotten so an individual's store of knowledge remained very limited.

It is not generally appreciated that we owe our vastly expanded knowledge to the invention or evolution of the written word. Imagine if everything you know had to be passed onto you verbally. Your store of knowledge would be very limited indeed. The written record changed that forever. Initially, knowledge carried by the written records was utilised by the select few that could read and write, the scribes, priests and suchlike. However, documents in whatever form were handwritten or carved or imprinted on clay tablets and were as scarce as hen's teeth, so the total store of knowledge remained very limited.

Then came the printing press and the possibility of mass producing many copies of the written word. Suddenly all of the accumulated knowledge of the world became accessible, at least in theory, to all who could read. Mankind's total store of knowledge exploded as ever more people wrote and disseminated books and others read them and accumulated this knowledge. The age of enlightenment wherein those who chose to could learn the wisdom of the ages became a possibility for the first time in human history. Books and paper ensured that the world's store of knowledge expanded exponentially.

We now are on the verge of an even greater quantum leap. We live in the information age wherein information can be stored digitally and distributed electronically. The Internet has provided us with a seemingly inexhaustible font of knowledge, although it must be admitted that much of what is there is of little value to anyone. However when one searches for a topic on a search engine such as Google and it searches 10 000 000 000 pages to find the information you are looking for then there is a realistic chance that you will find what you are looking for somewhere on the net and add it to your store of knowledge. This is but the beginning. In a move that is as significant as the invention of the printing press,

Google recently announced that it had committed to digitizing 25 million books and documents presently under the control of major libraries and make them freely available on the Internet. At last this vast store of knowledge will be available at the push of a mouse.

What has all this to do with a book that is primarily concerned with physics? One might imagine that now we have so much information at our disposal that the fundamentals of physics are not only known but also that they are cast in stone. If it were not so, someone would be on the "net" challenging conventional physics wisdom. A brief search on the net shows that literally thousands of people are completely disillusioned with the state of physics as it now exists. There are as many theories. The common thread is that everyone is searching for the fundamentals that would better explain the nature of matter and the forces of nature. It's a revolution against conventional physics that seems to have all but lost its sense of reality. If our mysterious hidden universe seems to be far-fetched, then the universe of conventional physics is more so. Conventional physics whose thought experiments far outweigh science fiction for the audacity of the concepts has taken a giant leap into the quasi-paranormal to its own detriment,

No, physics as known today is *not* cast in stone. Every new discovery brings more questions than answers, so we might be forgiven for thinking that something is wrong at the fundamental level. The fact is that in the field of modern physics, most information has got a very short sell-by date. What was cutting edge physics a year or two ago is out of date today. The touchstone of truth in physics has always been the discovery of simple solutions to complex problems. But all in all if we look at modern physics, there is absolutely no sign of convergence wherein every discovery would point to a sort of simplifying master theory. It just gets more complex all the time.

As one example of new discoveries producing more questions than answers, consider the argument that raged in the scientific community about whether there was enough matter to gravitationally slow down the expansion of the universe and perhaps cause it to collapse in on itself. The theory was driven, not by what scientists had observed, but by what they thought would be a neat solution to the origins of the universe. How nice it would be to say that the universe had no beginning or end but that it cycled between a state of singularity, a big bang, an expansion, and a collapse back to a singularity, and so on ad infinitum. Now it turns out that measurements have shown that the universe is not only expanding but that the expansion is accelerating! The inevitable conclusion is that the universe had a discrete beginning. There are no theories that even remotely explain what forces cause this acceleration and, for the time being, the scientific community is in disarray as it tries to figure out what causes this phenomenon. It's a major mystery.

It is said, not without good cause, that major progress in physics is built on the graves of physicists. Indeed it seems that historically, the greatest hindrance to progress in the sciences has always been the reluctance of established scientists to honestly examine alternative theories. This is very understandable. The scientist is dependent on his particular expertise for his very existence. He has built his reputation on knowing more than other lesser mortals about his discipline. His very reputation and livelihood are on the line. What if the very basis of his expertise were shown to be false? It would be a total disaster for most leading scientists. They therefore guard and defend their theories jealously and resist any competing theory especially if such theory emanates from outside the "establishment". I once contacted a leading physicist about the theory that will shortly be revealed in this book, giving enough critical information for him to be able to assess whether the theory had any merit. His response was to tell me to get a Masters degree in physics, proceed to a Doctorate and submit the theory to the critical examination by peers in the professional journals, "because this is the way it is done". He showed no interest whatever in the rather startling outcomes of a theory which if correct would impact seriously on the whole discipline of physics.

I am full of admiration for the brilliant minds of the scientists past and present, mainly because I now realise how the deck has been stacked against them all this time. In spite of the handicap of the rather shaky foundations of the discipline and having been led down many false trails, they have managed to come up with so many brilliant solutions to physics problems. My job has been much, much simpler. By casting off all the baggage of fundamental assumptions, a very simple (some might say naive) theory has evolved that enables us to view matter in a completely different and much simpler way. The prerequisite for appreciating the theories presented in this book is a completely open mind. The theories of some pretty big names in the physics world will come under scrutiny and some very well established theories will be challenged. Make your own assessment, free of the burden of preconceived ideas especially the idea that we know it all in the world of physics. Come and join the evolution of the mind as we try to find the God Particle and unravel the mysterious sub-universe.

CHAPTER 3

The What, How, and Why of Physics

Conventional Physics theory is brilliant at describing what happens in any physical process and how it comes about. It rarely says why thing happen. Emission of light from the atom, wave particle duality and the shape of the graph of the strong nuclear force are given as examples. Nobody is saying why these things are as they are.

Conventional physics has been brilliant at describing the What and How of physical phenomena. It rarely describes the Why. For example consider the phenomenon of light emission from the atom. Ignoring for the moment, the muddying influence of the quantum physics view of the process, conventional theory explains that if an atom has energy added to it the electrons jump to a higher energy state, absorbing energy as they do so. When the electron relaxes to a less energetic state, light is emitted. *What* has happened is that the atom has emitted light. *How* it has done so is by allowing the electron energy state to change. *Why* the atom should behave in this way or why this should result in the emission of a wave packet consisting of tens of thousands of laterally oscillating energy cycles (a photon of light), is simply never addressed.

Consider the other great mystery about this most common of phenomena. It can be shown by various experiments that light indeed consists of a wave train of electro-magnetic energy as described above. In other experiments, light apparently behaves as a particle with all the characteristics of matter. *What* has happened is that the photon is behaving like a particle. *How* it has happened is because the experiment was designed to detect particles. *Why* the photon exhibits this dual wave/ particle duality remains a mystery.

There are many more examples. Why *does* the Electron or the Proton exist as stable particles while the Neutron, in isolation, is not stable? What is the nature of the strong nuclear force that behaves in a totally erratic manner switching from repulsion to attraction and then disappearing completely as the distance from the centre of the nucleus increases? The graph of this force was discovered and confirmed experimentally. *Why* it is so odd has never been addressed.

These phenomena are not exactly new cutting edge discoveries. The experimental work was done many decades ago. And yet we seem to be no closer to answering these seemingly fundamental questions now than we were when the experiments took place. Physics has simply leap-frogged or ignored the problem and gone on to other more complex and exciting prospects.

If conventional physics is unable to explain these base phenomena then surely we are justified in believing that there may be a more fundamental level in the material world that answers the *why* questions. Conventional scientists even have a name for such a system. They call it the "hidden variable". This book is about just such a variable, addressing the underlying basis of all matter and competently answering the *why* questions of physics. It's a book about a hidden universe based on the God Particle.

Chapter 4

What We Don't Know

Contrary to popular belief there's an awful lot we don't know about Physics. The list of unknowns includes the fundamental mechanisms of gravity, light, the strong nuclear force, the electron, Protons, Neutrons, charge, missing matter in the universe, the expansion of the universe, big bang theory, Gamma Ray bursts, why Relativity and Quantum theory do not agree, how mass is converted to energy, the nature of Planck's Constant, or why it is impossible to give physical meaning to quantum events. Even proponents of quantum theory seem to be disquietened by it. The proposition is made that there is a sub-structure to the physical world that will enable us to describe the mechanisms underlying these phenomena.

Mystery No 1: Gravity

It is the most common of the forces of Nature. It keeps our planet just the right distance from the sun, it keeps us from flying off into space, it determines the flight path of every star in the universe and is universal in nature. There is no cubic centimeter in space that is not affected by gravity. We know enough about gravity to send a satellite into space, have it slingshot around Mercury to pick up speed, come back and slingshot around the Earth to pick up even more speed, and then send it on its way to Saturn some 1,2 billion km from earth. Then with exquisite precision pre-adjust the path of the craft so that it descends without further instruction onto the moving target of Titan, a moon of Saturn. By any measure this voyage must rank as the greatest feat of space navigation ever competed successfully by man. This complex navigation requires us to know how gravity works to a very high degree of precision. So we know all about gravity, right?

Not really. There is unsolved puzzle called The Pioneer Anomaly. Spacecraft Pioneer 10 and Pioneer 11 blasting off towards the edge of space at a constant 25000 miles per hour are actually slowing down contrary to the laws of gravitation Though we have now lost contact with these spacecraft, this anomaly, although very small, remains a mystery.

We know more or less how gravity works to a very small degree of error. What we don't know is why gravity works. What is the mechanism underlying gravity? Why do two bodies at a distance from each other apparently attract each other instantaneously without the possibility of a signal passing from one to the other? Ever since Newton applied his mind to the problem these things have remained unknown. The best conventional physics is able to come up with is that gravity like any other force-at-a-distance is a field. But then we have to ask what a field is other than lines on a piece of paper or a mathematical expression. There seems to be no answers to these vexing questions. Einstein, as usual, had something to say regarding gravity. He insisted that space be defined in terms of four dimensions, the fourth dimension being time. In terms of this (non-Euclidean) approach, the presence of matter distorts space geometry and this manifests itself as a gravitational force. This unproven theory seems to have little concrete meaning. Einstein died in 1955 still searching for the way to describe gravity and the rest of the physical universe in terms of this geometry.

Mystery No 2: Light

Conventional physics has it that when an electron changes orbit from a higher state of excitation to a lower state, a photon of light is emitted. Conversely, a photon can cause the electron to jump to the next possible orbit. The whole concept of light emission from the atom in conventional physics is problematic in the extreme. This situation is made even more uncertain if one considers the quantum model of the atom. Here the electron does not actually seem to orbit the nucleus and the best that could be said of it is that at any instant there is a probability that the electron will be somewhere in the vicinity. How then does the atom go about emitting light? There are simply no answers.

A second great puzzle about light is the apparent wave / particle duality of light. Sometimes the photon displays all the properties of a series of electromagnetic waves. At other times it displays all the properties of a particle. We simply don't know why. In this regard one of the simplest of experiments produces results that still defy explanation. It's Young's famous double slit experiment and its refined versions.

To summarize Young's experiment briefly, a ray of light is directed towards two narrow slits in a screen placed in front of a second screen or light sensitive plate. The light passing through both slits produces an interference pattern of dark and light bands due to the destructive interference and constructive reinforcement of the light waves, which travel slightly different distances on the different paths to the second screen. This experiment demonstrates the wave nature of light in a convincing fashion. If one of the slits is closed the pattern disappears.

The bizarre part of this experiment is that if the intensity of the light beam is reduced so that only one "photon" at a time hits the second screen and if both slits are open, the light dots (recorded sequentially and photographically on the second screen) summate to present an identical interference pattern! As soon as one slit is closed, the pattern of dots changes dramatically and the interference pattern is no longer to be found. Identical results are obtained if electrons are used instead of photons. This is a real puzzle. How can particles produce the same interference pattern if they go through the slits one at a time! Conventional answers to this problem are bizarre in the extreme. Realistically we simply have no explanation.

Mystery No 3: The Strong Nuclear Force

This is one of the most basic of the forces of nature. It holds the protons in the nucleus of atoms together in spite of the fact that they should repel each other because the have the same charge. It's a strange force that we can measure accurately. Very close to the centre of the nucleus this force is negative, i.e. repulsive. A bit further out it is positive, i.e. attractive, and a very short distance from the centre of the nucleus it suddenly disappears completely and suddenly. Though we know exactly how it behaves we have absolutely no idea why it behaves in such an odd manner.

Mystery No 4: The Electron

This is one of the fundamental particles of all matter. We know it exists as a particle and we know its mass to a high degree of precision. What we still do not understand is why the electron should choose to orbit a nucleus (classical model) or what in fact it does if the location of the electron can only be represented by a "probability cloud" surrounding the atom where greater probabilities are represented by higher cloud densities (quantum model). The quantum model doesn't even inform us what the electron is actually doing. A further confusion is that like the photon, the electron can either be a particle or a wave. We simply do not have any explanations for this apparent Jekyll and Hyde characteristic.

Mysteries 5, 6, and 7: Protons, Neutrons and Charge

Protons and Neutrons are fundamental components of atoms. They are very similar particles having slightly different masses. The main difference is that the Proton has a positive "charge" and the Neutron does not. We do not know why this is so nor do we know why a Neutron in isolation from the nucleus of an atom is unstable and decays into a Proton and an Electron. It just does, but there has to be a reason and a mechanism of conversion. Nor do we have any fundamental explanation of exactly what "charge" is! We can only measure its effect but we don't know what causes it.

Mystery No 8: Missing Matter

Estimates vary but physicists agree that between 90% and 95% of matter in the universe is "dark matter". There is every indication that it is there but we can't find it. We just don't know anything about it.

Mystery No 9: The Expanding Universe

It has been known for some time that the universe is expanding. We know this because measurements of light emanating from distant Galaxies are "red shifted" that is the normal spectrum of light has shifted towards the red end of the spectrum. This indicates that these galaxies are everywhere moving away from us at more than half the speed of light. Now it has been discovered that this expansion is accelerating! No one knows what is driving this accelerated expansion. Whatever force it is must be greater than the restraining force of gravity.

Mystery No 10: The Big Bang

If the outer reaches of the universe are receding from us at half the speed of light and if the speed is accelerating, then working backwards the initial speed of outward movement when the universe was created would have been very much less than that. Does the Big Bang become the Big Whimper? We just don't know.

Mystery No 11: Gamma Ray Bursts

There is a phenomenon that occurs with monotonous regularity about which we know little or nothing. Appearing out of space are intense emissions of Gamma ray radiation. Their spatial distribution is very even. They come from all directions and are very intense. The problem is that if, as is suggested, they come from very distant Galaxies, the amount of energy emitted would have to exceed the total energy of the universe to arrive here at the intensity observed. There has to be an alternative source for these bursts, but we simply do not know enough about them. A new satellite, the Swift, has been launched at a cost of about $160 million to find out more about these phenomena.

Mystery No 12: Relativity and Quantum Theory.

Here we have a real dilemma. Quantum theory is well established and is probably the most proven theory ever to be developed. Relativity, likewise, has been accepted as being true and there are many proofs of this. The problem is they do not agree! In the macro world relativity holds sway. In the micro world, Quantum theory is king. And yet there is no transition point from macro to micro. Both lay claim to the universality of their application. We simply have no idea how to unify these two great theories.

Mystery No 13: Conversion of Mass to Energy.

$E = mc^2$ is probably the most famous of all scientific equations. It states that energy is equal to mass times the square of the speed of light. Now the square of the speed of light is a very large number indeed that in effect says that very small amounts of matter can be converted to huge amounts of energy. One only has to look at an atomic submarine or an atom powered aircraft carrier that can travel 30 to 40 knots for years on end at the expense of a few kilograms of uranium to realise this. So where did Einstein get this equation from? Like most of Einstein's work, he took the work of others before him and built on it. This equation stems directly from Lorenz's mathematically derived transformation equations. Einstein merely put flesh onto the bones to point out that mass could be transformed into energy. It was the basis for the atomic bomb. The equation undoubtedly works but we don't understand why, when one splits an atom's nucleus, and the component parts weigh less than the original nucleus, then the missing mass becomes energy. We simply do not understand the mechanism of this transformation although we make use of it in every atomic bomb and atomic pile.

What We Don't Know

Mystery No 14: Planck's Constant

In 1900 Max Planck applied his mind to a problem that had been plaguing the mind of the leading physicists of the day. Experiments in measuring the temperature (frequency of emission) from a black body showed that the curve of energy against frequency was shaped like a skewed bell. Planck made the arbitrary assumption that the energy was emitted in packets or Quanta and derived the expression $E = h\nu$ which fitted the curve almost exactly. The little h in that equation would rock the physics world. It was the start of Quantum Theory. Undoubtedly the equation worked, but what, if anything, did the constant mean? Or what indeed did its dimension, Ergs x seconds mean? He didn't know, we don't know, and nobody seems to know. It's still an unsolved mystery.

Mystery No 15: Quantum Uncertainty

You might be forgiven if you are starting to think that if we don't know all these things, what do we know? The question becomes even more relevant as we progress into the strange world of Quantum Physics. The development of quantum theory and its outcomes is dealt with comprehensively in Chapter 7 and it is difficult in the extreme to summarize it in this short format. Suffice it to say that quantum physics and uncertainty theory leaves us almost bereft of any meaningful or visualisable models of the world around us.

The theory originated in the mind of Heisenberg who in a very powerful thought experiment reasoned that one could never determine the position and the momentum (manner of movement) of a particle simultaneously. This meant that for the first time in the history of physics here was something that could never be measured. Extreme interpretations of this principle then demands that in fact the particle does not *possess* the properties of momentum and position at the same time since what cannot be measured does not exist. It's either/or, not both. Ultimately by extending quantum theory to the extreme conclusion, merely thinking about a particle or looking at it is enough to alter it and prevent you from ever knowing how it was behaving. This extreme view was the creation of Bohr another heavyweight in particle physics and, like Heisenberg a physicist who pre-eminently looked to mathematics to solve his physics problems. Bohr explained the wave-particle duality by declaring that subatomic systems don't have either of their complementary states until they are observed. Once observed, the "thing" made up its mind whether to be a particle or a wave. This view came to be known as the "Copenhagen" interpretation of quantum mechanics, named for the city where Bohr and his colleagues operated. Bohr was also responsible for the Complimentarity Principle that stated that:

Evidence obtained under different experimental conditions cannot be comprehended within a single picture, but must be regarded as complementary in the sense that only the totality of the phenomena exhausts the possible information about the objects.

Once more the boundaries got smudged. In effect Bohr was saying that no single experiment could tell you what is happening in the world of physics.

The net result was to blunten physicists desire to discover the fundamentals of their craft. After all who in their right mind was going to challenge these giants in the world of physics? Better to go with the flow and pretend to understand the un-understandable. One is reminded of the story of the King's clothes.

Mystery No 16: The God Particle

Scientists around the world are assiduously trying to do mega-experiments to plug the fundamental gaps in their understanding of the universe. One of these gaps concerns the so-called Higgs Boson also nicknamed the "God particle" by Nobel laureate Leon Lederman. It was so named because its total pervasiveness yet complete un-detectability likened it to divinity. This Boson and its field are supposed to confer mass to all other particles. A huge and complex theory has evolved around the Higgs Boson. The problem is that it has never been discovered. Undaunted, the scientists compete with each other to see who can spend the most money to build the most powerful particle-colliders to discover what is essentially a totally theoretical particle. It's a huge extravagant gamble.

Conclusion

The above list of what we don't know would seem to indicate that in spite of the huge growth of knowledge of the physical world, physics is really in its infancy. Physics has simply leap-frogged the fundamentals. The reasons for this are to be found more in philosophy than in science. Niels Bohr was a real heavyweight in the physics discipline and his views about particle physics became established even over those of Einstein who, in turn, was deeply disturbed at the direction physics was taking. It is not to be thought that the proponents of modern quantum physics were altogether satisfied with their own theories. There was an innate uneasiness which is still felt today.

"Physics", Bohr declared, "tells us not about what *is*, but only what we can say to each other concerning the world." Bohr had abandoned the quest to understand Quantum Theory.

Heisenberg speaking about emission of light from the atom said "...but this is not supposed to happen with the electron: instead the frequency of vibration of the emitted light is said to lie somewhere between the orbital frequency before the mysterious jump and the orbital frequency after the jump. All this is sheer madness."

Another view attributed to Morris Kline puts it this way:
"Unfortunately, the course of twentieth-century science is departing farther and farther from 'common sense,' from intuitively accessible concepts, and from simple, physical pictures. More and more science is resorting to complicated mathematics for which the physical account is either incomplete or even inconsistent despite the fact that this account is real enough to design and produce atomic bombs."

Richard Feynman is credited with writing "I think it is safe to say that no one understands quantum mechanics."

Schrödinger described quantum mechanics as "a formal theory of frightening, indeed repulsive, abstractness and lack of visualisability

The renowned mathematician, John von Neumann was an adherent to the extreme view that claims that the universe does not exist until a human mind is there to observe it.

If such is the trend of thought of leading physicists, then pity us poor ordinary mortals! There are many more unknowns in the world of physics, but the above gives just a smattering of how undeveloped the discipline really is in terms of fundamentals. It seems as if the practitioners of physics have simple pressed on to deal with more complex and esoteric issues and forgotten about the basics. In this book we will be returning to fundamentals and try and solve these mysteries in a rational and realistic way. In doing so we will make the assumption that there is an underlying structure to all of matter – the mysterious God Particle and the hidden universe. We suspect that the physicist's response would be that we are assuming that there *is* some underlying mechanism for each phenomenon and that this is not necessarily true or that this covert universe is simply a figment of an overactive imagination. We believe that it is provable that there is indeed an underlying mechanism to each of these phenomena. Mechanism of course implies the use of classical physics and we will be calling on Newton's laws throughout this book.

Chapter 5

The Hidden Universe

The proposition is made that there is a substratum underlying physical universe, as we know it today. This substratum consists of an Aether-like Cosma consisting of primary particles called Zerons (God-particles) that collide and inter-react to form a resonant "perfect" fluid. This substratum cannot be detected by direct means and it constitutes a hidden universe that determines the behaviour of all matter and the prime forces of nature. Only two things are invariable in this hidden universe, the speed of light and time. Zeron theory will enable us to unravel much of what we don't know in an objectively real manner. All in all Zeron theory will produce a completely new alternative model of matter and the forces of Nature.

In the face of all the unknowns of physics, it seems to be clear that there must be a common thread to all these problems. In general physics has stopped at saying what happens and determining the rules or laws regarding how it occurs. In every case it stops short of describing *why* matter and the forces of nature do what they do. We believe that there is an undiscovered sub-stratum to the physical world, one which explains the "why" detail of the phenomena we experience. This book is about the proposition that there is just such a sub-stratum. This will provide the solutions to the unknowns of physics. We have named the prime particle on which this sub-stratum is based the Zeron and the theory that backs it the Zeron Theory.

In Zeron Theory we make one basic assumption. In the entire physical universe, there exists just one fundamental prime particle, the Zeron. All matter is composed of various arrangements of this particle and the forces of nature are determined by its properties. The existence of this substratum cannot be determined by direct measurement and we have therefore called this sub-structure the hidden universe.

The God Particle

We are surrounded by and are part of this hidden universe. We can't see it or feel it or construct an instrument to detect it. Yet this clandestine universe is real in every sense of the word. It determines the foundational structure and behavioral properties of all matter. Without it no atoms could exist and you and I would not be here. This universe is not only invisible, it is undetectable by direct measurement. In the language of the physicist, is therefore, at best, a virtual universe. At worst we can't measure it directly and what cannot be measured does not exist. And yet this hidden universe is as objectively real as the atoms and molecules that are us.

This all sounds so unlikely that you may be forgiven if you think at this stage that this book is purely speculative and that it can have no place in scientific literature. Be assured that, on the contrary, every step in the development of this theory has not simply been pulled out of the air, but is based solidly on the re-interpretation of the key physics experiments and theories of the past. There is so much circumstantial evidence for the existence of this hidden universe that it simply cannot be written off as being imaginary.

Before delving into detail it might be well to illustrate where we are going. Zeron Theory starts with the assumption of the existence of the Zeron. By re-interpreting the results of many of the key physics experiments of the past a model of the properties of this Zeron is deduced. A key part of the development is the outcome that the Zerons in free space would inter-react with each other to form a sort of Zeron atmosphere akin to the Aether of old but with very different properties. There are some surprising results. In the hidden universe there are only two invariables. The first invariable is the speed of light (c). The second invariable is Time (t). This flies in the face of conventional physics particularly that of Einstein. There are convincing arguments for why time should be invariable and not variable and, as we shall discover it is this assertion that restores objective reality to physics.

In this hidden universe the surprising outcome of invariable light-speed and time is that everything else is variable. Our standards of dimension and mass and all their derivatives are (as seen from a universal perspective) variables, changing incessantly, according to where we are in our orbit around the sun and where our sun is in the universe. Don't expect to be able to measure any of these strange phenomena in a direct manner. No direct measurement or detection is possible. And yet this variability is the real essence of reality.

Zeron theory will permit us to re-interpret many theories and solve many mysteries. We will discover what makes Gravity work, how the atom produces photons, why photons can be waves or particles and why Young's double slit experiment works with one electron or photon at a time. We will find out about the peculiar nature of the strong nuclear force. We will discover that Gamma ray bursts are short-lived manifestations of the Cosma into the visible world. By treating time as an invariable we shall be able to distinguish between what is observed and what is real Einstein's Relativity Theory and we will add a term to these equations thus removing the counter-intuitive outcomes. There will be harmony between Relativity and Quantum physics. We will be able to say why mass can be converted to energy and derive the $E=mc^2$ equation from a variety of sources. The meaning of Planck's constant will be revealed, and the mass of the Zeron will be deduced. Then we will discover the seemingly impossible, a Universal Frame of Reference. We will take a critical look at the key experiment that establishes Uncertainty Theory and produce enough evidence to negate both the experiment and the theory. Many other side issues will be critically examined. All in all Zeron theory will produce a completely new alternative model of matter and the forces of Nature. We will build this mysterious hidden universe particle by particle. Whether it is valid or not will be left to the reader to judge.

To lay the basis for the development of Zeron theory we need to look initially at the two opposing systems of Classical and Quantum physics.

Chapter 6

Classical Newton

Newton was the greatest scientific genius that ever lived. The laws of physics that he stated permit us to attach objectively real properties to matter and its properties. They permit us to ask the question, "What is the mechanism behind any phenomenon?" Newtonian laws constitute Classical Physics and this will become the cornerstone of Zeron Theory.

Many hold that Newton, physicist, scientist, philosopher, mathematician, astronomer and alchemist, was the greatest scientific genius of all time. A little known fact concerning his talents was that he was also an expert on the bible. He spent more time studying the bible than he spent on his scientific endeavors. Unfortunately his dissertations regarding the bible are less well known and most were fanatically destroyed by those who came into possession of them, regarding them as being heretical.

Born on Xmas day in 1642 his best-known work the Philosophiae Naturalis Principia Mathematica, or "Principia" as it became known, was published when he was 42 years old. It formed the foundation of physics for the next 350 years. Principia is generally acknowledged as the greatest scientific book ever written. It is ironical that but for the generosity of his friend Edmond Halley who encouraged him to publish it and financed the publishing costs, Principia might never have seen the light of day.

His accomplishments were astounding. He formulated the law of gravitation, the laws of motion, the law of conservation of energy, invented calculus, developed the binomial theorem (a key theorem in the development of mathematics), gave mathematical expression to Kepler's laws of planetary motion, correctly analysed the properties of light passing through a prism, constructed telescopes, and studied the speed of sound through air. Single-handedly he took physics into the modern era and was the originator of what is now called *classical* physics.

The main characteristic of classical physics is that all processes within it can be described as mechanistic. The question "what is the mechanism behind this phenomenon?" can always be answered in classical physics by applying Newtonian laws. These laws basically define the behaviour of real objects in a real world. If you push a thing it moves and barring frictional and other losses will keep on moving that way indefinitely until you push it in a different direction. If you keep pushing it accelerates, and if you double the weight you halve the acceleration. If you push something, you feel it resisting your push. It's simple, intuitive, and in the words of Einstein "objectively real".

By examining the hidden universe that underpins all physical phenomena we will be returning the discipline of physics back to the Newtonian principles and laws. Back from what? Back from the chaos of Quantum Physics.

Chapter 7

Quantum Chaos

We look at the history of the development of Quantum theory from Planck's quanta to the uncertainty theory of Heisenberg. The subtle slide from uncertainty (uncertain error of measurement) to indeterminacy (uncertain properties of matter) signaled the end of a deterministic universe in the minds of leading physicists. It remains so today. The extrapolation of uncertainty theory means that the operator's observation and even his awareness of the experiment changes its outcome. Uncertainty theory has led us into the quasi-paranormal. The solution to this dilemma is to be found in the hidden universe founded on the existence of a truly indivisible prime particle, the Zeron.

The discipline of physics underwent a major change between 1900 and 1930 when a number of leading physicists, feeding off each other's work, changed our concepts of space, time and matter. This is the age of quantum physics and uncertainty. Nothing is as it seems and all the rational laws of physics that were for so long considered to be part of nature are no more. It's a difficult world to understand, especially coming on the heels of a classical style model of the universe which was profoundly intuitive and which made sense to the average person like you and me.

The history of the development of Quantum Theory is as bizarre as the theory itself. It started with Max Planck. Born in 1858 into an academic family, Max Planck received his Doctoral degree at the University of Berlin at the tender age of 21. He was the Professor of Theoretical Physics at that university for some 38 years. His personal life was tragic. He lost a wife, two daughters and two sons, one of which was tortured to death by the Gestapo for suspected involvement in a plot to kill Hitler. Max plank a brilliant but broken man died in 1947. He left behind a legacy that has troubled the world of physics ever since.

Quantum Theory has its roots in Max Plank's suggestion in 1900 that light, x-rays and other radiation were not emitted at an arbitrary rate but could only be emitted from the atom in certain minimum packets that he called "quanta". He arrived at this rather unusual conclusion in the process of finding a formula to fit the well-established experimental graph for "black body" radiation. It was found from experiment that the curve of the observed rate of emission of radiation from hot bodies plotted against the wavelength of the radiation was shaped somewhat like a skewed bell. Wien and Raleigh two leading physicists of the day had tried unsuccessfully to derive a formula to fit the curve. By his own admission, Planck, without any supporting experimental evidence, simply guessed that the radiant energy was emitted in packets or quanta. His inspired guess had had no theoretical basis, but his hypothesis neatly resolved the problem of finding a formula to fit shape of the curve. It was moreover an elegantly simple formula. Had Planck made his mark on history on the strength of an inspired guess? Not really. He looked at the evidence and modeled a workable solution just as we will be doing here.

Planck opened a Pandora's Box. Even today we still feel the after-effects. Planck derived an equation that gave the energy of the radiation as a constant times the frequency of the radiation. He called the constant "h". He was at a loss to explain what this constant meant - if anything. Furthermore the constant was found to be extremely small and had the puzzling dimension of Ergs x Secs. Nobody since seems to have done any better in explaining this constant. The best explanation that anyone could come up with was that somehow this constant was an expression of the underlying "grain" of matter. And there the matter stood and still stands today 100 years later.

It was not until 1926 that Heisenberg formulated his famous Uncertainty Principle based on Planck's quanta. Heisenberg had a long distinguished career in Germany leading to the award of the Nobel Prize of 1932 for the development of quantum mechanics leading to the discovery of novel forms of hydrogen. Heisenberg was fundamentally a mathematician and whenever he was challenged in the physical field, he resorted to mathematics to find his solutions. He was therefore firmly in the camp of the theoretical physicists, abandoning reality if his equations so demanded. He is probably most famous for his Uncertainty Principle that was to set the world of physics back on its heels.

Quantum Chaos

During WW2 the allies knew that Hitler had ordered the development of the atom bomb and that German scientists including Heisenberg were working feverishly to develop this ultimate weapon. A tanker load of Deuterium (heavy water), essential for the German nuclear programme, was sunk by Allied bombers and their effort to produce the bomb before the end of the war was critically delayed. After the war the successful American multinational nuclear team was astounded to learn that the Germans were, in fact, nowhere near producing the atom bomb. They had all along been barking up the wrong nuclear tree. Perhaps it is a measure of the vulnerability of Heisenberg's mindset and theories that he headed up that unsuccessful attempt.

Heisenberg's the Uncertainty Principle says that if one wanted to predict the future position and velocity of a particle one could shine a light on the particle and measure it's position and momentum (the way it is moving) by means of the light scattered by the particle. The more accurate the measurement, the shorter the wavelengths of the light needed to be, and by definition, the shorter the wavelength the higher the energy of the light wave. In order to affect the particle as little as possible it was therefore necessary that as little light as possible was used in the experiment. But the smallest amount of light is one of Planck's "quanta". Heisenberg showed that the uncertainty in the position of the particle times the uncertainty in it's velocity times it's mass can never be smaller than a certain quantity which turned out to be Planck's constant. Moreover this constant representing the uncertainty in the position or momentum of a particle is independent of the method of measurement and is a true constant.

At first glance this proposition seems to be innocent enough, having little relevance to the real world. However within this proposition lies the seed of a revolution in scientific thinking. Here is one of the major points of departure in scientific investigation that has influenced the way in which we view matter. In simple terms the proposition means that we can never accurately determine both the position and momentum of any particle. In my own rather simple view this merely places a limit on the accuracy with which we can determine these parameters, but to the cognoscenti of the physics world it had far greater significance. Here for the first time was something that could not be measured no matter how accurate the apparatus. This caused a subtle but important shift in thinking.

If either of two related properties of a particle could not be measured, then surely the particle did not posses one of those properties, or at least not both at the same time? The particle would then posses a sort of Jekyll and Hyde character and which one it was at any particular time would depend on what was being detected. This was the ultimate mystery. Matter was no longer necessarily matter. For instance, matter could be a "waveform" or it could be a particle. A particle could be here or there or somewhere in-between. All was uncertain.

This extreme extrapolation of the hypothesis meant that "uncertainty" (an expression of the limits of accuracy) was equated to "indeterminacy" a far more general property implying that there is something very strange about particles that makes them immune from exact analysis. This indeterminacy was then stated to be an inescapable fundamental property of matter. The retreat from reality had begun.

This development signaled the end of a deterministic universe. It was, and is currently held, that if we cannot measure the state of a particle without disturbing it, thus making the measurement invalid, we certainly cannot predict its future behaviour. This in turn is extrapolated to mean, for example, that an electron simply does not posses both a position and a momentum simultaneously. There is no explanation of how this can come about or what it means in physical terms. (We shall see later that this peculiar proposition is very close to the truth but in a way that the originators of the theory never dreamed of!)

The argument proceeds that there is an inherent fuzziness in matter that does away with the notion that an electron or photon or whatever moves along a distinct trajectory or path or has a determinable position. A quantum particle can have a position or a momentum but not both at once. An outworking of this rather puzzling statement is that matter takes on a surreal and ghostly nature. Is the object that you are looking at really there as you see it or has the mere act of looking changed it so that you cannot be aware of what it was like before you looked at it? Uncertainty Principle makes that claim.

An even more extreme extrapolation suggests that a person's awareness and consciousness actually form part of the object being observed as, quantumly speaking, there is no clear dividing line between the observed and the observer! A further result of the

Uncertainty Principle is that one can never be sure, at any moment in time, of the status of an electron a neutron, a proton, or any other fundamental particle. It can exist as a particle or a wave function and the mere action of observation may be enough to change its status.

There is a fine philosophical (that dreaded word) point here. Does the fact that we can't measure the status of a particle without disturbing it indeed mean that the particle has an inherent built-in "uncertainty" property in it's undisturbed state or is it merely that we have no means of measure it without incurring a small but definable error? At the bottom line it is the answer to this question that decides whether we keep science in the realm of objective reality or ultimately place it in the realm of the quasi-paranormal.

Quantum mechanics, in spite of this element of uncertainty and unpredictability has been the most successful scientific tool in history, underlying for example the explanation of the behaviour of transistors and integrated circuits. It is also the basis of modern chemistry and biology. The argument is extremely strong therefore Quantum mechanics is a valid scientific discipline and, by extension, we have to conclude that uncertainty is indeed an inherent quality of matter. Einstein, in spite of having been awarded a Nobel Prize for an (unwitting) contribution to the yet-to-be-developed Quantum Theory remained resolutely opposed to the probabilistic nature of matter that the theory seemed to point to. He refused to believe that "God would choose to play dice with His creation".

So here we sit, surrounded by a universe that, by definition, defies rational explanation. Fortunately it is not the end of the line. The solution to this dilemma is to be found in the hidden universe founded on the existence of a truly indivisible prime particle.

Chapter 8

The Prime Particle

We introduce the ultimate prime particle called the Zeron. The properties of this particle are enumerated. Its properties and behaviour lead to the formation of a resonant "atmosphere" of Zerons called the Cosma. The Quark, the prime particle of conventional physics has so many different properties that it too would seem to be made up of sub particles.

The idea of a prime indivisible particle is not new. There have been many iterations of the idea of such a particle starting with that of Democritus some 2400 years ago. It is fascinating to see the close correspondence between Democritus' concept of the prime particle and the definition of the prime particle as envisioned in this book. Truly, there is nothing new under the sun. To Democritus, the prime particle was:

- Invisible because of their extremely small size
- Indivisible as their name indicates solid (no void inside)
- Eternal because they are perfect surrounded by an empty space (to explain their movement and changes in density)
- Having an infinite number of shapes (to explain the diversity observed in nature)

It is astounding to find that some 2400 years ago Democritus created a model so accurate that we are now about to revive in an almost identical form.

Like the big bang, we have to start somewhere with our new theory. We don't intend taking you through all the thought experiments and years of investigation that preceded this start. That would be boring indeed. But after many moons of careful and methodical investigation we found substantial evidence of the existence of a truly fundamental particle that we have named the Zeron.

The God Particle

Zeron theory rests solidly on the hypothesis that the Zeron is the ultimate prime and indivisible particle.

Could the Zeron really be the ultimate primary constituent of matter, a real God Particle? We believe it is. A Zeron is undetectable even by the most sensitive apparatus. In conventional physics terms, this is bad news. Anything we can't detect doesn't exist! We should stop right here. While it is true that no direct experiment can ever detect a Zeron, this same undetectable Zeron will be shown to be the basic elementary particle responsible for the existence of all matter and the moderator of all forces in nature. There is more than enough circumstantial evidence for the existence of the Zeron. The Zeron has the following properties:

- All Zerons are identical and have the same mass and properties but come in two varieties, fast and slow.

- Zerons are perfectly elastic.

- There are no frictional forces between Zerons.

- Zerons have inertial mass but zero gravitational mass.

- Fast-Zerons permeate all space and are in constant movement at a constant speed that is the square root of 2 times the speed of light ($\sqrt{2}c$).

- Zerons form a resonant atmosphere akin to the Aether by impacting on each other lossessly. We have called this atmosphere the Cosma.

The properties may seem at first glance to be somewhat puzzling. Why should these properties and be chosen and not perhaps some other properties? Surely this is simply a long-shot guess?

Not really. The properties presented here are the survivors of many assumptions made about the Zeron. More assumptions were rejected along the way than were kept, so the above represents a highly refined list that did not exist at the beginning of the study, but one that has evolved hand in hand with the theory.

The closest approach to a primary particle in conventional physics is perhaps the Quark. The fact that Quark comes in "flavors" of up, down, strange, charmed, bottom and top, each with the "colours" red, green and blue is indicative of the bewildering complexity of even this most basic of particles. No physical model, or meaning, is attached to any of these strangely named properties and the quark remains as mysterious now as when first hypothesized. What seems to be indicated however is that for so many properties to exist there has to be some sort of substructure.

With the advent of extremely powerful atom smashers the discovery of new subatomic particles is routine. They are generally characterized by their extremely short life span before they degenerate into more stable forms. These are artificially produced particles that seem to have little relevance in the real world. The worrying part is that as of yet there seems to be no convergence in sight that would point to some unifying theory. It's as if the more we look the more complex the problem becomes. It is with this background that we seek to introduce a new theory that points to just such a unification.

Chapter 9

The Cosma and the Aether

We create the model of the Cosma, a resonant "atmosphere" of Zerons constituting a "perfect fluid". Comparisons are made between the Aether and the Cosma and we look at the history of the Aether. We re-look at the Michaelson and Morley experiment to detect the Aether and give reasons why the result is invalid. Einstein pronounced that, for his purposes, he did not need an Aether bit never denied its existence. The properties of perfect fluids and their relevance to the Cosma are explained.

The inevitable consequence of Zerons rushing around at high velocity is that they will bang up against each other and rebound. As the rebound is between perfectly elastic particles and because there is no friction between them, the rebound particles retain all the energy of the initial colliding particles and the process goes on ad infinitum with undiminished speed, frequency and energy. The process is somewhat analogous to the behaviour of molecules in air that continuously inter-react and bounce off each other. The impact of the molecules on any surface is what constitutes air pressure. Likewise the inter-reacting Zerons form an "atmosphere". In fact air is, in some respects, like a Zeron atmosphere. We can't normally see it and it exerts a pressure on every surface although we are not normally aware of this.

As all the Zerons are traveling the same speed i.e. $\sqrt{2}$ x speed of light, each of the Zerons is colliding with its next-door neighbor with near the same frequency. In fact the Zerons get into step with each other and the Zeron atmosphere resonates. We have named this resonating "atmosphere" the Cosma. It is unseen and is undetectable by direct means, but nonetheless, is responsible for all the phenomena of physics. Now that is a pretty audacious statement but we will justify every aspect of it in the course of this book. Before we examine the Cosma in detail it is certain that comparisons will be drawn between the Cosma and the Aether and the similarities and differences need to be elucidated.

The God Particle

As seen in Chapter 7 some of the deductions arising from the Quantum Theory and the Uncertainty Principle cannot be easily reconciled. A convenient escape for scientists faced with physical phenomena they cannot explain is to find refuge in Bohr's definition of physics. Physics, he declared, "tells us not about what is, but what we can say to each other concerning the world." In similar vein is the mindset of "If we can't measure it either doesn't exist or there is no point in trying to explain it."

We live in a world that is dominated by those subscribing to Bohr-type philosophy, and it seems that we are poorer in our knowledge as a result. In strong contrast, the position is taken here that if the circumstantial evidence concerning the existence or behaviour of anything is strong enough even if it be "undetectable", it is unnecessary to measure it or quantify it in order to verify its existence.

The Aether was an early casualty of the "we don't need it, let's ignore it, it doesn't exist" philosophy. Aether became the big myth of physics. In the early days of theoretical physics it seemed to the great scientists of the day that it was inconceivable that light could travel in a perfect vacuum. Drawing an analogy from sound waves traveling in air it was postulated that there was an undetected and maybe even undetectable medium pervading all space. It was this medium which was the carrier of light waves. It was named the "Aether" (Ether). The existence of Aether was hotly debated for many years until ultimately Michaelson and Morley conducted experiments supposedly sensitive enough to detect this mysterious substance. If one assumed the existence of the Aether, then as the earth traveled through it on its voyage around the sun there should be some observable effect on light depending on whether the observation was made against the supposed Aether flow or at right angles to the direction of motion. They accordingly raced one half of a divided light beam head-on into the anticipated Aether "wind" and the other half was raced against the first half but at right angles to it.

DIAGRAM 1

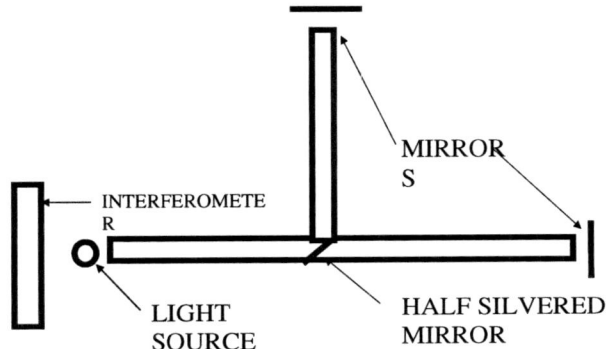

MICHAELSON AND MORLEY EXPERIMENT

A sensitive interferometer at the conjunction of the two beams detected no change in velocity or frequency of the two beams. The outcome was that there was no observable difference between the beam traveling in the direction of travel of the earth around the sun and the beam directed towards the sun. The apparatus was right, the math was right. It was therefore concluded that the Aether did not exist.

Fitzgerald and Lorenz (from whose work Einstein developed Relativity) thought differently. They suggested that there could have been a shortening of the measuring apparatus in the direction of Aether drift that would exactly compensate for any changes in the properties of the light beam. If this was true, the experiment proved nothing. The argument raged on. Fitzgerald and Lorenz were right and so in a twisted sort of sense was Einstein. Shortening indeed takes place. By means of Zeron theory, we shall show exactly *how* this shortening takes place. It will also be shown that Fitzgerald and Lorenz were more correct than Einstein. Now we are being a bit obscure, but all will become clear soon! Einstein, with his genius for lateral thinking, found a way to do his calculations without the necessity to include the concept of an Aether at all. Therefore it didn't matter for his purposes whether Aether existed or not. Furthermore Einstein's Special Theory of Relativity predicted exactly the foreshortening effect suggested by Lorenz. He concluded that it could never make any detectable difference whether a body moved through the Aether or not.

Einstein was right in that conclusion but wrong in the way he proceeded from that point.

Right here we are at the crossroads. This is precisely where physics goes down the road leading to incomprehensibility.

The proposition that it did not matter to Einstein's theories whether there was an Aether or not came, in time, to be interpreted that there was no Aether. Like so many obedient pupils following the master, physicists, then and now, neatly sidestepped the concept. We agree with Einstein's view of the non-detectability of the Aether, but hold that, whether detectable or not, the Aether plays a vital part in physics. Now the Aether is reinvented in a new guise with a new name and different properties, the Cosma. This will be the key to the re-establishment of an objectively real description of the forces of nature and of matter.

Something needs to be said about the properties of what are technically known as "perfect fluids". According to fluid mechanics, it is mathematically provable that objects move without resistance through a perfect fluid. In the perfect fluid there is no drag on a body moving through it, no vortices form and winged flight in such medium would be impossible.

It's the lack of friction between Zerons and the resulting zero viscosity that makes the Cosma a perfect fluid. Little did the creators of fluid mechanics realise that the perfect fluid was not merely a theoretical simplification to enable rather backward students such as I to understand what they were on about, but that the whole universe depends on and is immersed in just such a fluid!

A key property of perfect fluids is that while there is no drag on a moving body, resistance to motion is only experienced if the body is *accelerated* through the fluid. Constant velocity of an object through a perfect fluid produces no drag. Perfect fluids also have zero viscosity. To show that these unusual properties and effects do not only exist in the theoretical domain of hydraulics theory, we only need to look at liquid Helium II at 2.7 degrees K. This is a most remarkable liquid, which exhibits all the properties of a perfect fluid. It will flow through the finest of porous materials (apparently without resistance), has no surface tension, has an infinite specific heat, and even has the propensity, in defiance of the laws of gravity, to climb out of open containers, (a property of the fluid that has not yet been explained). Liquid Helium II at 2.7 degrees K is a real-life perfect fluid. It's a model for the Cosma.

Chapter 10

Any objections?

We analyse possible objections to the properties stated for the Zeron. The Zeron is the basic building block of matter and its properties determine all the prime forces of nature.

It is reasonably sure that what has been presented so far raises more questions than answers. The following FAQ may go some way to answering some of those questions.

Can anything, including the Zeron, travel faster than the speed of light?
A key component of Zeron theory is that Fast-Zerons travel at velocities exceeding the speed of light. To be precise they travel at 1.414... or the square root of 2 times the speed of light. The first question that must be answered is whether this does not fly in the face of Einstein's Relativity Theory. According to Einstein, infinite energy would be required for the acceleration of any body to the speed of light. By implication nothing traveling at below the speed of light can ever be accelerated to a velocity equal to or exceeding the speed of light. However, there appears to be no absolute prohibition on the existence of *ab-initio* particles traveling faster than the speed of light. In conventional physics the principle of superluminary (traveling faster than the speed of light) particles has long been accepted and at least one particle, the tachyon, is theorized as belonging to this category of objects. Zerons are indeed ab-initio particles, particles that came from the Big Bang nucleus.

Can anything be perfectly elastic and frictionless?
That these particles must be perfectly elastic and frictionless follows from the principle of conservation of energy. (Good old Newton!) When two Zerons collide, their combined energy is the same before and after the collision. There can be no losses as there is simply no mechanism for the loss to occur. The only possible way for a loss to occur would be by some form of radiated energy. However we shall see that at this fundamental level, radiated

energy, electromagnetic energy and all forms of energy transference are manifestations of the kinetic energy of Zerons themselves. There can be no underlying mechanism at some more fundamental level to transfer energy away from a collision between Zerons. This *is* the fundamental level. Without a mechanism for losses to occur the Zerons *have* to be perfectly elastic. The same argument applies to friction. There is no more basic mechanism that could carry away energy losses that might occur as a result of the sliding contact of one Zeron on another. Contact between Zerons is frictionless.

Can anything have inertial mass and not gravitational mass?

The general Theory of Relativity assumes that there is an exact equivalence between inertial mass and gravitational mass. This has been tested to 1 part in 100,000,000,000. The fact that this equivalence has been the subject of much experimentation, and that it has now been determined to such a degree of accuracy, is perhaps a reflection of the fact that some did not feel comfortable with either the General Theory of Relativity or with the equivalence of the two ways of determining mass. This equivalence is best illustrated by an example.

If you travel in a lift, as it accelerates upwards, you feel the additional "pull" which has the effect of apparently increasing your weight. If you were to be placed in a windowless capsule and accelerated at constant rate, there would be no experiment you could perform to tell you whether you were being accelerated or were being attracted by gravity. The two things are indistinguishable. Furthermore it appears that no form of matter is immune to either the pull of gravity or the pull arising from acceleration. Is this not indicative of the connection between gravity and inertia? And yet there seem to have been no attempts to analyze or explain this connection. How odd!

All matter is affected by gravity. Well not quite. There is one exception. We shall see that while Zerons have inertial mass, they are unlike other particles in that they are unaffected by gravity. They have zero gravitational mass. The reason for this is that the Zeron itself is the prime cause of gravitational force. We won't leave this statement unexplained. The mechanism of gravity is described in detail in the relevant chapter. Suffice it to say at this point that once more we are looking at the fundamental level, gravity being the secondary outworking of the behaviour of Zerons.

If Zerons permeate all space why have they not been detected?

Zerons lie at the very root of all matter. As will be shown throughout this book, the circumstantial evidence of their existence is compelling and in that sense they are indeed detectable. Zerons per se are not directly detectable. In the conventional scientific sense therefore, the Zeron is a virtual particle in that it cannot be detected by particle detectors. That this is not a far-flung idea we must refer to the "Tachyon" of conventional physics which is also hypothesized as having a velocity greater than light and which has an "imaginary" mass (whatever that means!). The hypothesizers of the Tachyon may have been closer than they imagined to resolving all their problems! Pity they didn't know about the Zeron!

How small is the Zeron?

Very small indeed. In fact it is the smallest thing imaginable this side of nothing. It is fair to say that the Zeron constitutes an entirely new class of matter beyond the wildest imaginings of even the most daring science fiction writer. How small is small? Consider a hydrogen atom. This has and one proton at its core and whizzing around it in some or other way is one electron. It is the simplest of all atoms. One could fit 25 million hydrogen atoms in the length of this dash - . If we look at the next level of sub-atomic particles, and place a marble on the ground to represent the proton in the nucleus, the electron would be about half a kilometer away and would weigh approximately 1/1800 the mass of the proton. Yes it is true. What appears to be an atom is nothing but a vast space with a speck of material at its core and an even smaller speck doing an odd sort of jig around the nucleus. The Zeron is even more bizarre. It would take about 1 000 000 000 000 000 (10^{15}) Zerons to make 1 electron!

This is the basic stuff of space and matter. It holds atoms together, is responsible for all the prime forces of nature, gravity, the strong nuclear force, the weak nuclear force the electrostatic force and magnetism. It is the basic component of matter and moderator of all forces.

Chapter 11

Matter Mysteries

We give a short history of the development of models of atoms from the plum pudding model through the nuclear/electron orbital model to the fuzzy quantum model. We give a preview of the strange characteristics of a Zeron based universe. We introduce a new field called the quantum field which is an integral part of any body but which extends well beyond what we have previously demarcated as the material limits of the body. An analysis of the properties of the Zeron enables us to develop a novel model of matter. Confirmation of the Zeron model is obtained by analysis of the graph of the strong nuclear force. Further confirmation by examining the graph of nuclear density. We show exactly why when two nuclei coalesce or an atom is split there is an absorption or emission of the binding energy. This sows the seeds for an independent derivation of the $E= mc^2$ equation.

The history of the giants of the scientific world who started to unravel the mysteries of the atom is a fascinating one. The story starts with Dalton as early as 1805. He hypothesized an atom that was a constituent part of compounds. He found that elements combined with each other in proportions that could be expressed as whole integers. Dalton the Quaker was delighted to find this general relationship which he described as a "direct example of the regularity and simplicity generally observed in the laws of nature". Would that we could say the same thing today. Zeron theory will reinforce the opinion that the laws of nature are indeed simple.

The fact that elements combined in fixed proportions led to the concept that this could only happen if there was a combination of discrete particles of matter. These particles were called atoms, but at first no one had any idea how small they were. Experiments with combination of gases led to the astounding answer. At first there was great skepticism not only in the scientific community, but when the discovery became public knowledge, the general opinion was that Dalton had lost his marbles (or atoms). The minuteness of the atom was hard enough to comprehend but then there was another surprise when the electron was discovered. The electron was an order of magnitude smaller than the atom! This was *really* incomprehensible.

Early models of the atom were like a plum pudding with the electrons studded on the surface. Then came the greatest surprise of all. The atom was almost completely empty! Deep down in the core of the atom was a tiny speck of matter. Orbiting this nucleus at a relatively large distance from it were one or more electrons, but the rest was empty space! To those of us who have been taught about and now feel comfortable with this model of the atom, it is hard to conceive the incredulity that this discovery produced at the time.

The atom that was portrayed contained a nucleus around which orbited the small negatively charged electron. No doubt the model was derived from the relatively recent discovery of the details of solar system with the planets orbiting the sun. Much time and effort were expended on this model to make it fit observations. The sizes and masses of the component parts of the nucleus were determined. Spectroscopy, observation of the products of radioactive decay and more recently, the examination of the products of very high-energy particle collisions in massive accelerators, have all contributed to our knowledge of components of the atom. While this model has been refined, especially by quantum theory, one thing has remained intact in the process. The atom is still considered to consist of a nucleus with a fixed number of electrons moving in some or other manner around the nucleus.

Quantum Mechanics, the Uncertainty Principle and careful measurement have radically altered the finer details of the atom. It is now depicted as a "fuzzy" nucleus surrounded by a "fuzzy" electron cloud. There is only a mathematical probability that the nucleus will be found in any particular position within the general volume allocated to the nucleus. The same applies to the electron shell. In pictorial representations of the quantum atom, darker areas of the fuzzy looking nucleus or electron shell indicate a higher probability of the particle being in that particular region. Lighter areas represent lower probabilities. There is absolutely no indication of how the nucleus or the electrons are moving or why they do so. It's all very vague. For a discipline that should be giving us precise answers, this is strange indeed!

DIAGRAM 2

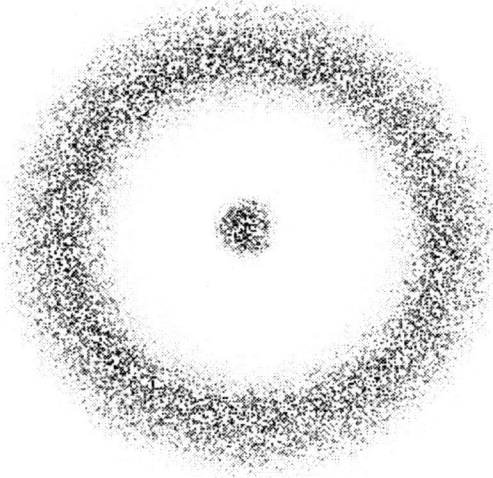

Our senses tell us that in the real macro world, matter is very solid in spite of the "nothingness" of the atom. We cannot see through a plate of steel. If we pick it up we can feel its weight. We have to apply appreciable force to dent it or bend it. It doesn't suddenly disappear as we watch it. It has every appearance of being unchanging, solid and real. Yet, in a way, what our sensory inputs tell us about the steel plate does not reflect sub-atomic reality. But however strange the conventional model of matter may be, the Zeron model is, in a sense, even stranger.

According to Zeron theory, way below the subatomic level, the steel plate has a peculiar and ghostly existence. The plate contains very little matter indeed and in fact it is has a rather nebulous nature. The miracle is that it continues to exist at all, for it is only because of the reliability of statistical laws that it does not simply disintegrate and disappear before our eyes! If all this sounds strange it is because at this sub level there is a very different universe, a hidden, invisible, undetectable sub-universe of great complexity consisting of an infinite number of interacting resonance patterns. In this sub-universe *nothing* is solid. Matter is merely a ghostly energy field, tenuous, transparent and existing

The God Particle

only in a statistical sort of way. Nothing exists for any appreciable time without completely exchanging the base particles from which it is made. All matter is, in that sense, a transient entity wafting through an invisible "atmosphere" of undetectable Zerons. Another strange outcome of Zeron theory is that the objects around us extend way beyond what we see or detect. They have a series of invisible resonance bands diminishing in strength according to the distance from the object, and theoretically at least, this compound "halo" extends way out into space. Yet these undetectable bands are as much an integral part of the object we are observing as the visible object.

Did you know that you are a shape shifter? Our very bodies and everything around us become shape distorted in different ways as we are whirled through space on spaceship Earth. Further with very few exceptions, what we now acknowledge as "physical constants" are not constant at all. Our standards of length and mass and all their derivatives (as seen from a universal perspective) are variables, changing incessantly according to where we are in our orbit around the sun and where our sun is in the universe. Don't expect to be able to measure any of these strange phenomena in a direct manner. No direct measurement or detection is possible. We are talking about a hidden universe but one that is the very essence of reality.

Such are just some of the strange phenomena revealed by Zeron theory. Such, by extension, are the properties of the "real" universe, the one we perceive with our senses! It is not to be thought that we are developing some sort of mystical or paranormal explanation of matter and the universe. On the contrary, every model created in Zeron theory is solidly based on classical physics - more than can be claimed for modern physics theory. Moreover, those parts of currently accepted physics theory that do not lead to "objectively real" descriptions of matter are firmly rejected.

We have said enough about what Zeron theory *will* show. Its time we got going with the models. As with all Zeron theory, these models are easily visualisable.

The first part of Zeron theory is that all matter is composed of Zerons and only Zerons.

It is convenient to visualize Zerons as perfectly elastic spherical particles. We immediately call on Newton to invoke the law of conservation of momentum. Two Zerons moving towards each other on a collision course will, in a totally elastic collision, rebound off each other without loss of energy. If the path of the Zerons is mutually directed towards the centers of the Zerons, the Zerons collide and are repelled at the same velocity but in the opposite direction. Where there are grazing incidences, the same classical reasoning applies but the Zerons are deflected at angles depending on the degree to which they graze each other. The important principle is that the total energy of each Zeron pair is unchanged because there is no more fundamental mechanism by which energy loss can take place. There *is* no other energy form other than the energy intrinsic in the movement of Zerons. The movement of Zerons is energy at its most basic level.

It is therefore fundamental that once Zerons are in motion there is also no mechanism by which the Zerons can be "persuaded" to stay in contact with each other and they continue in their random motion between impacts with undiminished velocity. (While this statement is generally true, we will see that special cases of resonant behaviour can cause Zerons to become sub-luminary. More of that later.)

It is fundamental that the closer the Zerons are to each other the less distance the Zeron has to travel before it impacts on its neighbor. Because the Zerons have the same velocity before and after each collision, smaller inter-Zeron distances mean higher the vibrational frequencies within the Zeron population. Higher vibrational frequencies are perceived as higher energy states.

If the energy level within any population of fast-Zerons is held constant, they get into step and all assume identical vibration frequencies. This resonant "atmosphere" of resonant fast-Zerons constitutes the model of the Cosma.

Having modeled a Cosma, let's go on to model matter. Imagine for a moment that a number of Zerons, for whatever reason, are found to be in close proximity to each other to form an agglomeration. This agglomeration will be bombarded from all sides by the fast-Zerons in the Cosma. There will be a tendency for the random impact forces imposed by fast-Zerons to break up the agglomeration and return the components of the agglomeration to the random movement of Zerons within the Cosma by simply knocking off vulnerable Zerons. This is analogous to a professional style "break" in a game of Snooker where the outside balls are removed from the body of the colored ball formation by means of a grazing impact by the white ball. After many assaults by the white ball on the on the formation, the balls are left spread around the table in a random formation. Thus it is with sub-nuclear particles artificially manufactured in atom smashers. They have very short life spans and they quickly disappear, emitting a burst of energy as they do so. There is no conventional explanation of what is happening here except to invoke Einstein's $E=mc^2$ equation to explain the production of energy from matter. Wave a magic wand and it happens! The Zeron theory equivalent is very straightforward. The Zerons comprising any agglomeration less than critical size are simply returned to the Cosma. The resonant frequency of the Cosma in the immediate vicinity increases and we perceive this as increased energy levels.

The process of the return of ordered systems of Zerons to the randomness of the Cosma lies at the root of the well-recognized principle of the natural increase of Entropy of any system. According to this principle all systems naturally become less ordered with time. As Zerons move from an ordered state to a less ordered state, the Entropy of the system is said to increase. In the above example we have neat visualisable model of what is going on when a system increases its Entropy.

We *still* do not have a model of matter. Relax! It's about to be constructed. If we consider a body made up of a large enough agglomeration of Zerons, a stable particle may result. Imagine an agglomeration of Zerons, as before, but one which is composed of a very large number of Zerons so that the size of the agglomeration is many orders of magnitude greater than a Zeron. All Zerons in the Cosma are vibrating at the same frequency and are therefore in resonance with each other but are traveling in random directions between collisions. The large agglomeration is impacted on all sides by normal (perpendicular to the surface) and grazing impacts of the Zeron population around it. If the agglomeration is large enough, for every fast-Zeron impact, there is a component of that impact force that is normal to the surface of the agglomeration. This normal component is, by definition, directed towards the centre of the agglomeration. The sum of these force components results in an overall "Zeron pressure" directed towards the centre of the agglomeration. This pressure forces the individual Zeron components of the agglomeration towards each other. Such an agglomeration held together by Zeron pressure may then constitute a stable particle that we recognise as matter.

There are, however, a few problems with this model. If a number of Zerons within the resonating Cosma are in some way forced towards each other to form an agglomeration, the frequency of Zeron impact within this agglomeration would rise exponentially. This would create an ever-increasing internal force of repulsion. For an agglomeration to be stable the externally imposed Zeron forces must exceed the internal repulsive forces.

By definition the repulsive force generated by Zerons at small inter-Zeron distances, such as would be found in a compressed agglomeration of high velocity fast-Zerons, would be greater than the external Zeron pressure generated by the further-apart fast-Zerons. No stable particle could result because the inside-to-out forces keeping the agglomerated Zerons apart would always be greater than the outside-to-in forces imposed by Zeron pressure from the Cosma. However if the Zerons in the agglomeration were to travel at lower velocities than those in the Cosma, a stable agglomeration becomes a possibility. This is the key feature that makes agglomerations stable. The Zerons in the agglomeration move at lower velocities than that of the fast-Zerons in the Cosma. Now for the first time we have a model that more or less works. Matter becomes an agglomeration of Slow-Zerons held together by the Cosma's pressure. This leads to a fundamental definition of matter.

Matter is defined as any agglomeration of slow-Zerons.

We still have quite a long journey ahead to model matter satisfactorily. There are all sorts of interesting resonance patterns and these determine the physical properties of the atom.

To reiterate, all atomic nucleons are composed of agglomerations of Slow-Zerons held together by Zeron pressure generated by the surrounding Cosma. "Held together" does not however mean that these Slow-Zerons are in contact with each other. Even at the centre of the nucleus the Zerons are moving, resonating against each other to create a repulsive force that holds them some distance apart. Now if all this sounds rather fanciful it would be good to seek some sort of justification for the theory. What proof is there to justify this description of matter?

The Strong Force

It is instructive to examine the generalized graph of the strong nuclear force plotted against the distance from the centre of an atom's nucleus. This well-known graph was determined experimentally many moons ago. The shape of the graph (thick line) indicates that there is a "hard core" of repulsion near the centre of the nucleus. This changes to an attractive force some distance from the centre. The attractive force diminishes to zero further from the centre. Ever since its discovery the shape of this graph has always been a bit of a mystery. What sort of force changes from repulsive to attractive and then suddenly disappears a small distance from the nucleus? One might expect at least that the attractive force would decrease exponentially and simply taper off as the distance from the centre of the nucleus increases. But no. It disappears completely a short distance from the centre of the nucleus! This is strange indeed! Zeron theory to the rescue.

The peculiar and hitherto unexplained shape of the graph can be clarified if the graph is resolved into two components of force as shown below.

Matter Mysteries

DIAGRAM 3

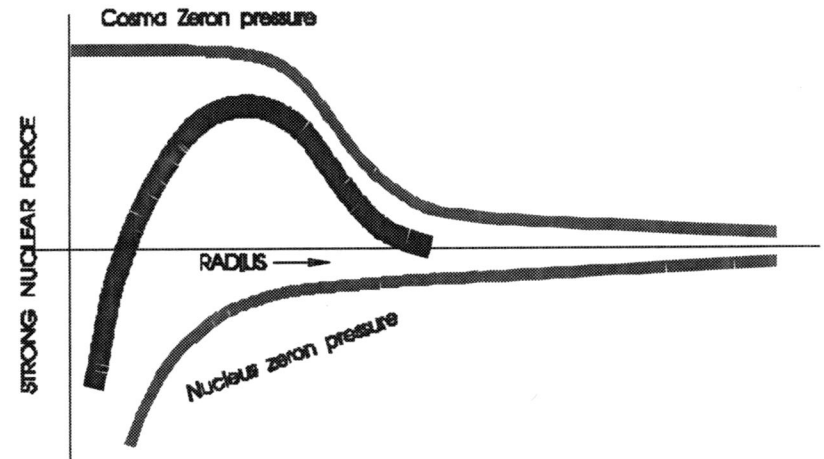

GRAPH OF STRONG NUCLEAR FORCE

The heavy line represents the generally accepted, experimentally determined curve for the Strong Nuclear force plotted against radius from the centre of the nucleus. We have resolved this graph into two different components. Below the x-axis is our new curve representing the repulsive force generated by the resonating Zerons within the nucleus. As the inter-Zeron distance reduces towards the center of the nucleus, the force increases rapidly. In fact, the force increases exponentially. The shape of the curve is typical of exponential curves.

Above the x-axis is the second graph. This is the external force induced on a Zeron agglomeration by Zeron "pressure". At the outer edge of the nucleus where the nucleus is fuzzy, there are just about as many impacts from in to out as from out to in. Almost as many but not quite. There is a shielding effect due to the presence of the nucleus. The inward force starts from a low value at the outside edge of the nucleus, where the inward pressure is mitigated by the outward pressure of the remaining fast-Zerons in the "fuzzy" outer layers of the nucleus. (More about the "fuzzy" later). The inward force increases towards the centre of the agglomeration as the "shielding effect" of the nucleus comes increasingly into effect. The inward force reaches a limiting value near the core of the nucleus and reaches this peak value at the radius of the "hard" core of the nucleus. Here there are effectively no more in-to-out random fast-Zeron impact forces to work against the out-to-in pressure. The inward force then levels off. This is depicted in the graph above the x-axis.

The sum of these two curves gives the conventionally recognized graph of the strong nuclear force plotted against radius.

Add the two curves and it is immediately apparent why the force goes from repulsion to attraction. It is because the bottom line has a bigger value than the top line. The reversal of the force occurs when the opposite applies. It also becomes clear why the force suddenly disappears. A short distance from the centre of the nucleus, the values of the graphs above the horizontal axis and below, cancel out and the resultant force is zero. At last there is a lucid explanation for the puzzling shape of the graph of the strong nuclear force.

Now about the "fuzzy" bit. The nucleus is not homogeneous in density. Indeed that it is not so is shown in the graph below. This too is an experimental graph obtained in neutron scattering experiments. The Zeron interpretation of this graph is as follows. At the core of the nucleus there is an agglomeration of Slow-Zerons held very close to each other (this core has been named the nucloid). The graph of density of the nucleus against radius shown below indicates that the compression effect on the nucleus is progressive so that the transition from the relatively "hard" nuclear core to the fast-Zerons of the Cosma results in a progressively less dense "mantle" of resonating Zerons around the core of the nucleus. Conventional physics, while providing the experimental evidence has no such rational explanation.

DIAGRAM 4

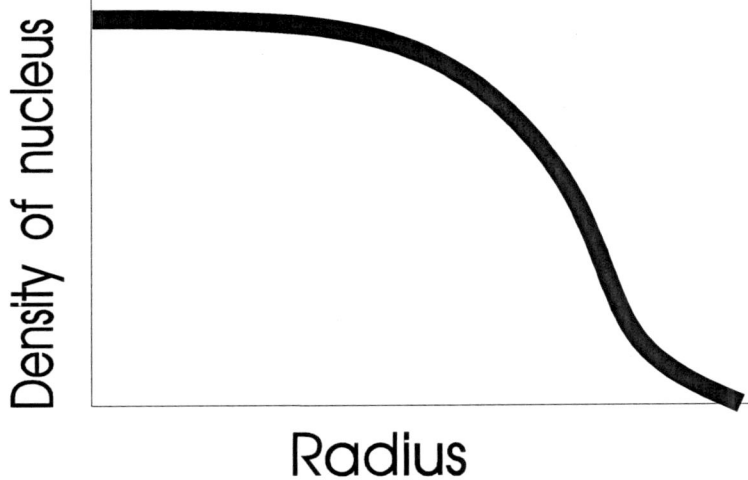

NUCLEAR DENSITY GRAPH

There therefore seems to be more than a little justification for this Zeron model. In summary, the Zeron model of a simple nucleus is one of a particle of matter with a "hard" nucloid held together by Zeron pressure, surrounded by a mantle of decreasing density in which there is ever increasing fast-Zeron activity.

If all this seems to be a bit complicated, take a deep breath, because we're by no means through with the atom. All we have is a fairly simplistic model of the nucleus of an atom. We will see that the model needs further modification to take account of resonance effects *within* the nucleus. These resonances even occur within the hard core of the nucleus.

How, you might ask, do agglomerations of Slow-Zerons come about? To ask what conditions result in Slow-Zeron agglomerations in a high velocity Zeron Cosma is to ask how matter is created. Clearly if Zerons are perfectly elastic, fast-Zerons in the Cosma maintain their initial superluminary velocity, no matter what. Slow-Zerons of which matter is made can only exist as ab-initio particles formed at the moment of creation. There is at first glance apparently no natural mechanism that can result in this exceptional grouping of Slow-Zerons in the Cosma. There is, however one exception. Resonance can cause the faster-than-light Fast-Zerons in the Cosma to become sub-luminary. It will be shown that a special case of resonance is at the root of the existence of the electron particle. Furthermore, resonance within the nucleus is the key factor that determines whether it will be stable or not.

A distinction must be made between vibration and resonance. In the vibration mode, a certain energy level is required to maintain a vibration at any particular intensity. In resonance mode this *same* energy can produce far greater intensities due to the fact that the vibrations build on and amplify each other's movement. The amplitude therefore grows until limited by some other factor. It will be seen that the nucleus of the atom and the Cosma interact with each other in a resonant manner to produce stable particles.

We now examine in more detail what happens when two nucleons coalesce. After coalescing, Zeron pressure will keep the new nucleus intact and roughly spheroidal, but the total numbers of Zerons provided by the two nucleons supply more than the correct number of Zerons required for resonance. They form instead, a spheroidal nucleus that vibrates but does not resonate. The reason for this is that though the nucleus may theoretically vibrate at any frequency, it can only resonate at frequencies compatible with that of the enveloping Cosma. It is the impacts from Zerons in the Cosma that sets up the nuclear vibration in the first place, but if there are too many Zerons in the newly formed nucleus, the Zerons on the surface of the nucleus don't get into step. To get into step the circumference of the nucleus has to be such that the tangential component of the Zeron vibrations on the surface of the nucleus set up a self-reinforcing, standing wave configuration. When this happens, the nucleus starts to resonate in sympathy with the Cosma. Should the circumference be too large, or too small, no resonance takes place. The Zerons in the outer mantle of the non-resonating nucleus become vulnerable to the effects of impact by the Zerons in the surrounding Cosma. The agglomeration will lose Zerons to the Cosma.

The above is a rather static model of a highly dynamic situation. Perhaps a more realistic way of looking at this process would be to recognise that there is a continuous interchange of Zerons between the Cosma and the nucleus. In a non-resonating and therefore unstable nucleus, the rates of absorption and ejection of Zerons is uneven. More Zerons are thrown off from the nucleus than are absorbed, until the nucleus is left with the requisite number of Zerons and the nucleus starts to resonate. The correct number of "resonance protected" Zerons is now in what has become a stable nucleus. Resonance protection occurs when the outer mantle of the nucleus generates a standing wave pattern in response to the effects of the resonant Cosma enveloping it.

In the immediate vicinity of the new nucleus the Zerons that are thrown off from the non-resonating nucleus increase the Zeron population, and thus the frequency of the Cosma in this region increases. This is interpreted by an observer as a release of energy. As a result of the ejection of Zerons, the mass of the new nucleus is less than the sum of the masses of the two original nucleons. In conventional physics it is said that this difference in mass is turned into energy. This is called the "binding energy" of the nucleus. Binding energy is quantified by Einstein's $E=mc^2$ equation, where E is the binding energy, m the loss of mass, and c the speed of light. We will come back to this Einsteinian equation many times for this is really one of the most universal expressions. The question remains, though, why does $E=mc^2$? We will show later how this equation can be derived from Zeron theory.

In the mean time we seem to have made some real progress. For the first time we have a fundamental description and rational explanation of the processes involved in this well-known phenomenon. We may now state a further proposition: The binding energy expressed as mass is always an exact integer times the mass of a Zeron and the total binding energy represents the total kinetic energy of the ejected Zerons. Do you start to see the connection between mass and energy?

Chapter 12

When is an Electron?

A resonating nucleus in a resonating Cosma leads to the formation of a Zeron Electron shell. There are no electron particles in the Zeron Electron shell! For the first time there is a meaningful explanation of the de Broglie wavelength for an atom. Electrons form at concentration nodes within the electron shell and $E=mc^2$ applies. These nodes form and reform in a dynamic fashion. Electrons only form in experiments designed to detect electrons!

Here we lay the foundation for a fairly radical departure from conventional physics. We find some really surprising things about the electron.

As we have seen above, the nucleus of the atom is held together by the radially inward component of the force of impact of Fast-Zerons on the nucleus - a sort of Zeron "pressure". However there is a limit to such pressure so that even at the heart of the nucleus the repulsive force generated by its resonating Slow-Zerons is strong enough to keep these Zerons a small distance apart. The Zerons in the core therefore continuously collide with each other and form a resonating core or nucloid. This nucloid is enveloped by a resonating mantle of decreasing density. For nuclei larger than that of the hydrogen atom the nucleus takes the form of a nucloid enveloped by resonance shells similar in character to the Zeron-electron shells that surround the nucleus. The resonating frequency within the nucleonic shells is, however, much higher. These nucleonic shells are close enough to the nucloid or central "hard" core to be considered as an integral part of the nucleus. For the purposes of the following section, the nucleus may simply be considered to be a resonating agglomeration of Slow-Zerons within a resonating Cosma.

DIAGRAM 5

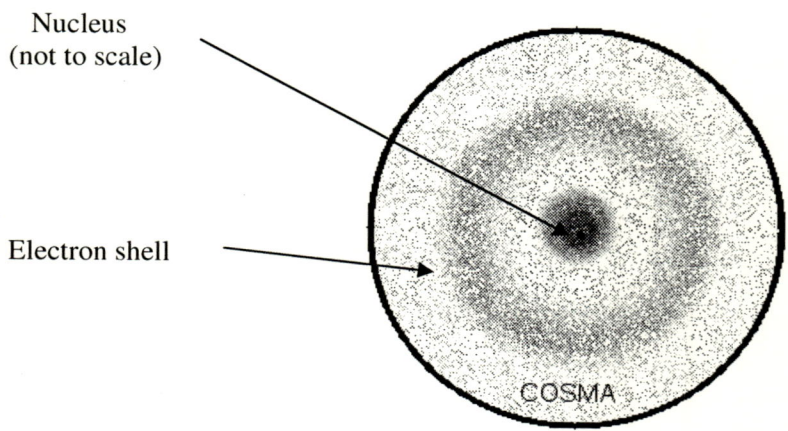

Nucleus (not to scale)

Electron shell

COSMA

We now have two resonant systems, the resonating nucleus, and the resonant Cosma surrounding it. Like the 3-D equivalent of a vibrating point in a pool of water, the resonant nucleus produces spheroidal, outwardly traveling waves of Zeron momentum in the Cosma. The momentum is carried forward by the impact of Zeron on Zeron in the form of an outwardly expanding sphere of impact driven waves. The Cosma provides a resonating environment through which these waves travel. The two resonating systems interact to produce shells of destructive interference and constructive reinforcement. This is the mechanism that forms what we have called the Zeron electron shell structure. It is the same mechanism that creates the nucleonic shell structure.

For any particular fast-Zeron energy in the Cosma and for a particular nucleus there exists a specific stable concentration of Zerons in spherical shell form at a constant distance from the nucleus. If one were to measure the Zeron population density along a radius it would show that at some distance from the nucleus there would be a narrow band in which there is a very high concentration of Zerons. This concentration is the spherical Zeron electron shell.

When is an Electron?

Now we have a nucleus complete with its resonant shell or shells. It is a stable entity. But where are the electrons? Read on for the intriguing answer.

The Zerons within the shell, being radially constrained by the resonant phenomenon, are unable to move much in the radial direction and therefore have a predominantly tangential motion. This resonating shell is stable for two reasons. Firstly because of its high energy level, the shell has the energy to withstand the incessant bombardment from fast-Zerons from the Cosma. Furthermore, within the shell itself, a tangential standing wave pattern is formed further adding to its protective resonant nature.

This resonant electron shell is the "quantum electron shell" of conventional physics. In quantum theory there is a high (but not exclusive) probability of finding the electron in this shell than anywhere else. The Zeron theory model shows a striking similarity to the Quantum physics model, but in sharp contrast, provides us with a visualisable and lucid picture of the processes involved.

There is one major difference between the Quantum model and the Zeron model. The difference is that, in the Zeron-electron shell, there are no electrons! It will be shown that the Zeron 'electron' shell is the breeding ground for yet-to-be-formed electrons, but in the stable state, no electrons exist anywhere! We must beg your patience while we examine the shell structure in more detail before explaining this rather startling statement.

Any agglomeration of Zerons will disintegrate under the assault of Zerons from the Cosma. The only protection is if the surface is in a resonant state and this applies to both the nucleus of the atom and its shells. The shells form as a result of the interaction of the nucleus and the Cosma but will achieve resonance and stability only at specific radii. The reason for this is that tangential standing waves must be formed in the Zeron shell before resonance can exist within it. This resonance is achieved when an integral number of the tangential Zeron waves fits exactly into the circumference of the shell. The waves then become self-reinforcing and self-destructing in a tangential direction and the requisite standing wave pattern is created.

The amplitude and energy level of the standing wave increase dramatically due to this self-reinforcement. The average wavelength of the standing wave in the tangential direction is determined by the combination of the energy level of the Cosma and the mode of vibration of the nucleus. The mode of vibration in turn varies from nucleus to nucleus depending on how many nucleons there are in the nucleus. This gives rise to different shell patterns.

Conventional physics explanation of this same phenomenon is rather circuitous and hinges on the still unexplained particle / wave duality. There is no real explanation of where the electron is to be found, or whether it is an electron or a "wave". There is however one striking agreement between the Zeron model and the conventional physics model. De Broglie and Bohr hypothesized that the electron "orbit" would be an exact multiple of what has become known as the "de Broglie wavelength" for the electron. This has an exact parallel in the standing wave wavelength in the Zeron-electron shell. Whereas there has never been a rational explanation for what the de Broglie wavelength of an atom actually is, the Zeron model, by contrast, presents a concrete and easily visualisable model.

When is an Electron?

DIAGRAM 6

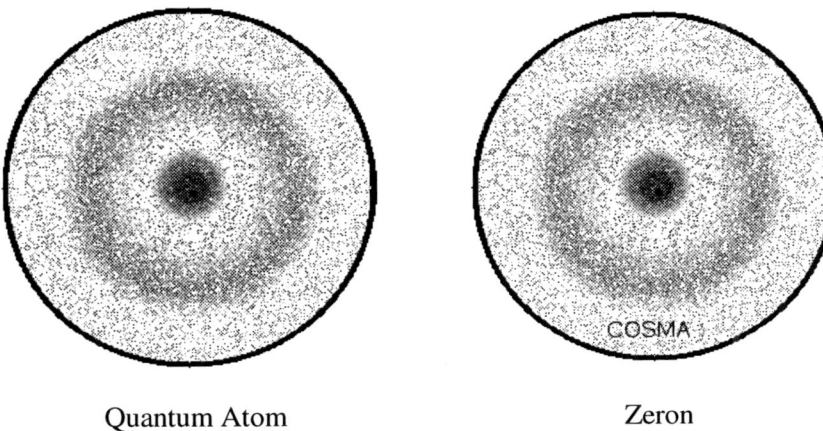

Quantum Atom

Darker areas = Higher probability of location

Zeron

Darker areas = Higher concentration of Zerons

If we look at graphic representations of Quantum model of the atom and those of the Zeron model we find that they are identical in appearance but not in meaning. On the left is the conventional quantum representation. Darker areas represent higher probabilities of the electron being found there. One of the unexplained mysteries of conventional physics is what causes the electron to behave in this rather odd semi random manner. What drives it? What determines this pattern of probability? Do I hear a deafening silence?

On the right is the Zeron model. Here we have an extremely simple model. We are looking at a graphic representation of Zeron population densities. Higher Zeron densities are found in the nucleus and in the Zeron electron shell. Makes sense doesn't it? The transition from one model to the other is accomplished by substituting "Zeron population density" for the Quantum Theory's electron "location probability" function. So the conventional scientists got it almost right- except for one major difference. There is no electron in the Zeron model! No electrons in the atom? Surely not! The electron is one of the foundational particles of physics. Does that mean that the Zeron model is incorrect? By no means! There are further models in store that will provide a realistic model to confirm this position.

Before we leave this simple model of the Zeron electron shell, we need to note a secondary effect within the shell. The prime effect within the shell is one of the transfer of Zeron momentum from Zeron to Zeron in a randomly orbital manner. The effect within the shell is therefore primarily the movement of momentum rather than the movement of Zerons. However each time the Zeron momentum transfer completes one orbit of the shell in any particular plane, there is a translational movement of one "pitch" (equal to the inter-Zeron distance.) of the Zerons themselves. Therefore on each orbit the Zerons in any chosen "momentum train" move one pitch. As a result of momentum transfer there is a slower but significant translational/orbital movement of the whole Zeron population within the shell.

Stepping back a while, we saw that the shell can only exist if it is in a state of resonance. Resonance in turn implies a standing wave formation with resonance peaks. The shell is of course not a two dimensional entity but is a three dimensional spheroidal object. The standing wave is generated in (potentially) an infinite number of planes each having its own resonance peaks. In practice, the resonance peaks in the shell achieve stability with the formation of small islands of peak resonance or resonance nodes. These nodes repel each other and thus form a stable pattern of nodes equidistant from one another and distributed equally over the entire surface area of the shell.

DIAGRAM 7

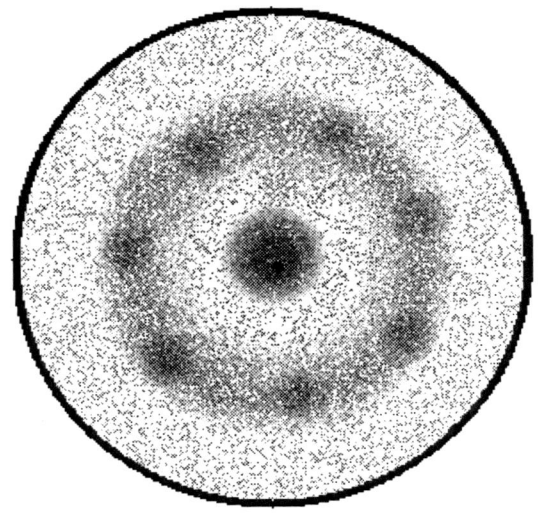

Two Dimensional Cross-section of atom showing
node concentrations.

The number of nodes depends on the type of nucleus and the radius of the shell. These nodes are not stationary; neither do they circulate in any particular plane. However, the superposition of translational movement of Zerons on the nodal pattern ensures that the positions of these nodes are not static. The result is that these high resonance nodes appear and disappear at different points in the shell in an apparently random manner but in such a fashion that the statistical mean position is one of equidistant but moving nodes. It is ironical to note that the 3-D representation of the above Zeron atom would mimic the original "plum-pudding" model of the atom! Maybe the early physicists were cleverer that we think. If we now put all the parts of the model together we land up with the following description of the Zeron electron shell:

The electron shell may be depicted as a resonating spheroidal shell of Zerons distinguishable from the surrounding Cosma by its higher Zeron density, having small equally distributed higher density nodes that build and disappear in a random manner.

There is an easy way to visualize this. Put water into a shallow container and subject it to vibration. Individual resonance nodes can be seen to form for a short time and disappear as adjacent resonance nodes form. If you increase the intensity of vibration sufficiently, you will see droplets of water being ejected from the surface at the nodes. This ejection of drops of water is the visualisable model of the formation of electrons.

This is where things get really interesting. We are about to confirm that there is no such thing as the electron in all but the most exceptional circumstances!

We made the statement that there are no electrons in our Zeron electron shell. The reason for this is that the shell is sort of pregnant. The electron is yet to be born. Like a human birth the electron emanates from a "bump"! Because of resonance and the standing wave formation the Zerons in the shell, bump-nodes composed of concentrations of fast-Zerons become the breeding ground for the rare electron. The birth of the electron can be visualized as the formation of a "droplet" of fast-Zerons pulled out of a node in the shell by one or other physical process to which the shell is exposed. Once isolated from the shell, the electron becomes an independent free entity. However under normal circumstances, there are no electrons in the "electron" shell of the atom!

When the right conditions exist, the agglomeration of Zerons is pulled from a resonance node in the shell just like the droplet of water leaving a vibrating water surface. Once the electron has left the shell, the shell contains too few Zerons to sustain resonance and the shell collapses or alters its resonance pattern. The atom is then considered to be "ionized".

When is an Electron?

DIAGRAM 8

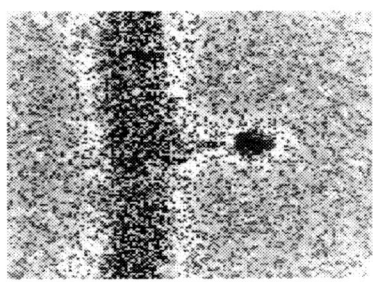

Electron

"bleeding" off an electron shell

The agglomeration of Zerons emanating from the Zeron-electron shell is unique in that it is one of a very few particles that is composed of fast-Zerons. This extremely small agglomeration is recognized by the observer as the electron particle. It has an extremely short life span and is only found in specific man-made experiments. More of this later when we look at electricity.

If we now look at the energy levels within the shell after the ejection of an electron, we find that the shell assumes a lower energy level and is left in a non-resonant and therefore unstable condition. The reason for this is as follows. When the shell was still in its fully resonant condition, the resonance resulted in an increased Zeron population density within the shell as compared to that in the surrounding Cosma. By definition therefore, the resonant electron shell constituted a high-energy region as compared to the energy of the surrounding Cosma. The formation of the electron particle is accomplished at the expense of part of this additional energy and the shell collapses back to the lower energy levels. Einstein's $E=mc^2$ applies to the formation of the electron (matter) from the energy of the resonant electron shell (kinetic energy of the Zerons). We will be coming back to this important link between Zerons and Einstein's equation because here is the first glimmering of what underlies this famous equation.

Under what circumstances will an electron agglomerate? The electron particle will form at a resonance node of the electron shell as soon as the right sort of interference phenomena influences the shell. One may then ask what sort of phenomena these are. From the scientific and experimental point of view the answer is surprisingly simple.

Any experiment designed to detect electrons creates, de facto, precisely those conditions necessary to cause the formation of an electron.

The electron particle by its very nature is a more ordered system than a resonance wave system and the particle mode is therefore the less favored order according to the second law of thermodynamics. It follows therefore that the electron will only be found in particulate form if the resonant state in the node is transformed into a particle when compelled to by an outside influence. In all other cases:

The natural state of the electron is the Zeron-electron shell waveform. This waveform is devoid of any detectable matter.

It may be that nature has played a huge practical joke on us in that the electron as a particle may not exist at all other than in the apparatus built to detect such particles! This proposition would agree well with the Quantum Theory principle that in describing matter whether wave or particle, we need to specify the apparatus as part of the quantum system. Build apparatus to detect electrons and electrons will be formed. Build apparatus to detect a waveform and guess what you will find! There is even a case to be made that the electron does not exist in nature at all! It might in fact be a creation of mankind that has created apparatus to manufacture, direct, measure and detect the electron particle.

The formation of the electron starts to explain the puzzle of particle wave duality. We will be examining this duality in much more detail in the following chapter.

Chapter 13

Physics Puzzles

We give a description of Young's double slit experiment that showed the interference patterns caused by light waves. We describe the physicist's dilemma when he gets the same result by sending photons or electrons through the slots one at a time where clearly there is no second photon to inter-react with. We describe Alain Aspect's critical experiment "proving" faster than light communication between photons. Einstein's opposition to the quantum explanation of this and similarly related phenomena leads to the EPR thought experiment. An account of famous but bizarre thought experiments including parallel universes and Schrödinger's cat.

Young's double slit experiment

As long ago as 1801 a scientist named Young performed a simple experiment, the further development of which, causes controversy even today. Between a light source and a screen, Young inserted another opaque screen in which there were two parallel slits. Light waves going through either of the slits have slightly different path lengths to different points on the second screen

DIAGRAM 9

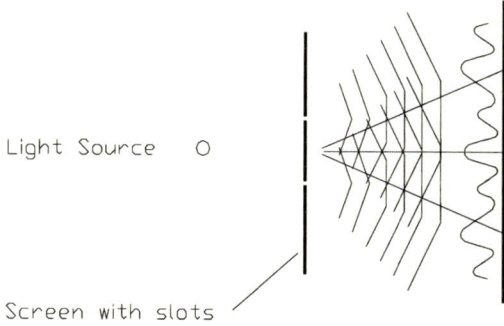

Young's Double Slit Experiment

The result is the appearance of an interference pattern on the second screen consisting of light and dark bands. This is due to the fact that the light waves passing through each of the slits get out of phase due to the different path lengths and they constructively reinforce or destructively interfere with each other. If one slit is closed the interference pattern disappears. This phenomenon is as the result of the wave nature of light. If there were no waves, "out of phase" effects could not occur. Young thus proved conclusively the wave nature of light.

The strange thing is that one gets exactly the same accumulative effect if one sends electrons or light "particles" (photons) through the slits *one at a time*. The particles appear to co-operate in some mysterious way to produce exactly the same pattern of interference! If one slit is closed the effect disappears! The quantum explanation for this puzzling behaviour is that the Uncertainty Principle demands that each electron/photon has an infinity of paths through the slits and they encode information about each other. How this encodation could possibly take place is not explained.

Nor is Richard Feynman's theory explained whereby each electron "travels by every possible path" to produce this interference pattern. At the end of the day, only one electron/photon travels at a time and has no other particle to interact with to produce interference. The only conclusion permissible is that with two slits open, the electron passes through both slits at the same time! This is typical of the counter-intuitive if not nonsensical outcomes of the uncertainty principal. There are many others.

The single photon version of the Young experiment is extrapolated as confirming the Uncertainty Principle. Flowing out of this Uncertainty Principle is the indeterminate universe. In this universe nothing is measurable, nothing is certain. All depends on the roll of some cosmic dice. So disturbing was the concept of an indeterminate universe that Einstein, in co-operation with Podolski and Rosen, created the following thought experiment (EPR experiment) to refute the Uncertainty Principle.

The EPR thought experiment.

The experiment addresses the proposition that it is not possible to know both the position and the momentum of a particle, one of the foundational propositions of the Uncertainty Principle.

DIAGRAM 10

v_1 . m_1

EPR EXPERIMENT

The thought experiment is as follows:

A single stationary particle explodes into two equal parts, which fly off in opposite directions. According to the principle of conservation of momentum, each part has equal momentum but in opposite directions. By measuring the momentum of one and the position of the other it would be possible to deduce both the momentum and position of either of the particles without violating the Quantum prohibition of knowing both parameters simultaneously. Moreover the measurements could take place when the particles were light years apart thus eliminating the possibility that any sort of signal could be sent between them. In Einstein's mind the universe was once more deterministic! One is tempted to wonder how many times the universe has altered its status due to the machinations of physicists! Fortunately for us the universe goes its own sweet way, regardless.

While Niels Bohr, chief interpreter of quantum Physics, disagreed with the EPR experiment and presented persuasive arguments to support his own point of view, the experimental confirmation of the uncertainty principle was to await Alain Aspect's experiment some half century after the EPR paper was presented.

Aspects experiment

Aspect's experiment depended on the powerful mathematical theorem produced by John Bell in 1965. Bell investigated the correlations that could exist between the results of measurements carried out simultaneously on two separated particles such as in the EPR experiment. He was able to establish a strict theoretical limit on the possible level of correlation for the simultaneous measurement of the properties of the two particles. Thus if an experiment could be devised to perform measurements in two isolated systems and if Bell's limit was exceeded, then Bohr's view of a Quantum universe would prevail in the sense that there would be some sort of co-operation or conspiracy between the systems. If Bell's limit was not exceeded then the "objectively real" universe of Einstein would prevail. The crucial experiment was devised and conducted by a team headed up by Alain Aspect.

DIAGRAM 11

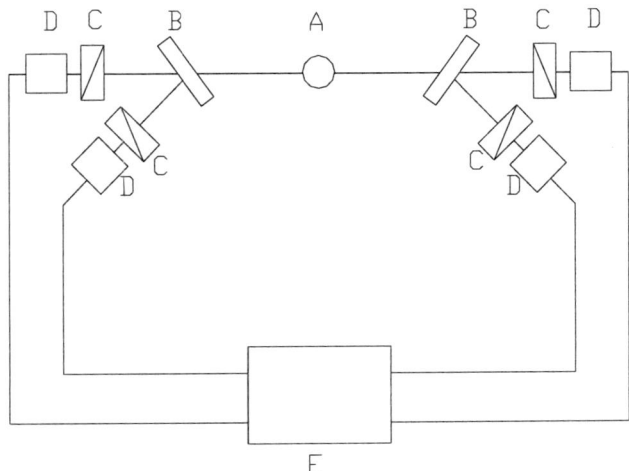

A Light Source B Optical Switches C Polarizers
D Photo multipliers E Electronic Coincidence Monitor

Aspect raced two photons of light along separate but identical paths. By using resonating optical switches he was able to delay the "decision" as to which of two polarizers each of the two photons would be directed to until the last possible moment. The arrival of the photons beyond the polarizers was monitored electronically. The apparatus was constructed in such a fashion that it was impossible for a speed-of-light signal to be sent from one photon to the other in time to affect the result. In essence the experiment was a sophisticated version of the double-slit experiment with the optical switching doing the equivalent of closing and opening one of the slits. The difference is that the slit equivalents were 15 meters apart. The results of a large number of photon tracking experiments confirmed that Bell's limit had been exceeded! This was a revolutionary outcome. In some way the series of two independent photons in two independent systems had co-operated in spite of the fact that no signal could have passed between them! Spooky stuff indeed!

The consequences of this experiment on scientific thought were profound. Bohr's position that it is meaningless to ascribe a complete set of attributes to some quantum object prior to measurement became the accepted dogma in the scientific world. Measure it and it changes! More bizarre thought experiments were to follow.

Quantum thought experiments

A sealed box contains a free quantum particle. Because no observation has been made the particle may be anywhere within the box the particle is represented by a "quantum mechanical wave" spread throughout the box. Were we to peer into the box through a peephole the wave would collapse and we could observe the position of the particle. Before we peek, however, an impenetrable screen is placed down the middle of the box. After the placement of the screen, it seems obvious that the particle can only be in one half of the box. Not so, according to quantum physics! Until we make an actual observation by peering into the box, the "quantum mechanical wave" still fills both sides of the box and the particle may be on either side. If we now make the observation and find that the particle is in the half of the box we are observing, the quantum wave in the other half of the box on the other side of the impenetrable barrier would collapse at the moment of observation even although it is completely isolated. If we observe no particle it is because the quantum wave on our side has collapsed and the particle has appeared in the other half of the box. If that nonsensical proposition seems to be far fetched, stick around, there is more to come! If now the two sides of the box were to be moved away form each other even to a very great distance, the same thing would happen.

The quantum wave in the remote box, perhaps light years away, would collapse the moment we made the observation in the local box and the particle might well suddenly appear in the remote box! Clearly danger signals were starting to appear. There was no lucid explanation of what was happening. The results of Aspect's crucial experiment were starting to bite.

Scientists now found themselves in a theoretical world that defied rational explanation. This was, and still is, particularly true in the sub-atomic region where stranger and stranger theories abound as physicists battle with these seemingly intractable problems.

Even more bizarre thought experiments were to be concocted. The famous branched universe theory was created in order to solve such paradoxes. If the scientist is faced with two alternatives, each legitimate, each scientifically proven and each being mutually exclusive, he is faced with an intractable problem. Each time a set of circumstances exists presenting such an unsolvable paradox, there can only be one answer and the answer must be legitimate throughout the universe. A way out of the dilemma is to have two universes. That way each of the solutions can exist in one universe and everyone is happy! This, believe it or not, is precisely the basis of the famous or perhaps infamous, "branched universe" theory. Whenever such a conflicting set of circumstances arises, the universe branches and in this way each branch may satisfy its own set of criteria and the paradox is avoided! So you see, in order that these paradoxes might not rock the scientific boat it has been decreed that you be cloned into hundreds if not hundreds of thousands of parallel universes! Quat paradoxes, tot clones.

Just as bizarre is Schrödinger's live cat / dead cat paradox. A theoretical experiment is set up in which a cat is sealed in a box with a piece of apparatus which contains a very small piece of radioactive material. At any specific time, the radioactive material may or may not decay. If it does decay, the apparatus will kill the cat. There is a peephole at one end of the box with a shutter. According to quantum physics, before observation, there is a superposition of two states. If the first state prevails we have a live cat and if the second prevails we have a dead cat. However the superposition of two states means we have a dead/live cat! It is only when we open the peephole and observe what is happening that one of the two states prevails and we find either a live or dead cat. This paradox arises from the notion that according to quantum physics one cannot draw a quantum line between the observed and the observer and the observation is the critical part of the experiment. Until you peek, the poor old universe simply cannot make up its mind!

If you think that these mental machinations are the result of hallucinations of some demented scientific soul that has partaken of something he should not have, think again. These were serious thought experiments set up by serious and competent scientists! The strange thought experiments arise due to completely intractable problems solvable in no other way. If, in the final analysis, the thrust of scientific endeavor in this field leads to theories that result in such nonsensical solutions, dare we say that something in the state of Denmark stinketh (and it's not only the dead cat)? Should we not look to see whether either Quantum Theory or Relativity or both have not been fully developed or are based on deeply hidden false assumptions that prohibit agreement between them?

One has to believe that something is missing in the theory or interpretation of either or both of these theories. Some scientists have suggested rather tentatively that underlying all these strange quantum phenomena is the "hidden variable", a sub-system producing the phenomena, but operating totally clandestinely. Zeron theory proposes just such a hidden variable. The hidden variable is the behaviour of Zeron-based matter.

Chapter 14

Throwing Light on Physics

We reveal the Zeron/Cosma basis of Newton's laws. For the first time there is a physical explanation of what causes inertia. We redefine a "field" in Zeron terms leading, to a physical meaning of a field. We introduce a new field called the Quantum field and show the importance of this field in debunking indeterminacy. We model the mechanics of light emission from the atom and seek confirmation by examining the graph of intensity of light emission from the atom. We give physical meaning to the Young's double slit experiment and show how single photons can produce the same interference effect. We explain the particle/wave duality of light. The conclusion is drawn that there is no such thing as a photon particle, only a "machine gun" effect that appears to be a particle. "Photons" will only be detected in experiments specifically designed to detect these "particles". We give physical meaning to the Heisenberg Uncertainty Principle in which an electron or photon possess momentum or position at any instant but not both and remove the uncertainty. We critically re- evaluate Aspects experiment, introducing the quantum field into the equation, and show why the results were not valid. We examine the second leg of Uncertainty Theory, namely the polarizing experiment, and show why results supporting uncertainty are invalid. If uncertainty theory is shown to be invalid, the universe can be considered to have been returned to a rational and objectively real status.

To recap briefly, as and as an introduction to this chapter on Newtonian principles, we restate a description of the new uncertain physics: "An extraordinary development in twentieth century science has been the putting to rest of the Newtonian 'truths' of absolute space and time, the unalterable objectivity of such phenomena as momentum and position, and the belief of an absolutely deterministic universe." Beautifully written, but fortunately not true. This section of the book examines Newtonian physics and returns it to its rightful "truth" position.

The Cosma is akin to a perfect fluid. Fluid mechanics tells us that there is no resistance to the motion of a body through a perfect fluid provided the body is traveling at constant velocity. Similarly, any body traveling at constant velocity through the Cosma experiences no retardation effect.

This phenomenon lies at the root of Newton's first law that states that if the net force on a body is zero, then the acceleration of the body is zero, and the object moves with constant velocity. Conversely a body in motion will continue traveling at constant velocity (no retarding effect) if no external force is applied to it. Thus Newton's first law describes exactly the behaviour of a body moving through a perfect fluid and by analogy through the Cosma. The behaviour of Zerons in the Cosma underlies Newton's first law.

Acceleration is a different matter. As we have stated previously, fluid mechanics tells us that if we accelerate a body in a perfect fluid, there is resistance to the movement. Consider what happens to an atom, accelerated in a perfect fluid such as the Cosma. Acceleration produces a force that opposes the force causing the acceleration. What is the root of such a force? Well if we accelerate an atom in the Cosma, two things happen. Firstly there is a bunching of Zerons ahead of the nucleus of the atom. Secondly, the nucleus starts to absorb more Zerons than it ejects and the Zeron count of the nucleus increases. This is because the bunching of Zerons ahead of an accelerating atom has the effect that there are a supernormal number of Zeron impacts on the upstream side of the nucleus compared to that on the downstream side. In a stable (constant velocity or stationary) situation, for every Zeron absorbed by the nucleus, a Zeron is ejected. The unbalance of the bunching means an abnormal number of Zeron impacts and a greater than normal absorption of Zerons into the nucleus. Do you get the impression that the nucleus would increase in mass? Dead right! But it's not quite as simple as that. The model doesn't, for instance, explain why the mass increase effect increases dramatically as the atom is accelerated up to near the speed of light. It's not a linear increase. We'll come back to this important effect shortly but in the mean time let us look at the forces on the accelerating atom.

The net result of acceleration on the atom is more Zeron impacts in front than behind resulting in a force that works against the direction of acceleration of the atom. Zerons act on each atom in the body. This effect is therefore greater for greater masses and is in direct proportion to the mass of the body. The macro result of this effect is that a differential Zeron pressure is applied to the body as a whole. This results in a force that opposes the acceleration. Does this sound familiar?

The phenomenon underlies Newton's second law that states that the force F is equal to the mass of the body times its acceleration. The equation is $F = ma$. In more common terms, if we try to accelerate a body it is as if there is a *force* resisting the acceleration.

We recognise this force as inertia.

We have now confirmed Newton's second law and have derived inertia from a real fundamental, the kinetic energy of the Zeron. There can be no more basic fundamental and inertia must therefore be considered to be a force in its own right and one of the basic forces of nature. In Zeron theory this force joins the list of the other prime forces, the strong force, the weak force, the electromagnetic force and gravity. Historically, the understanding of inertia has been fudged by its link to gravity. If you are thinking that the very close equivalence between gravity and inertia might lead to an understanding of gravity you are absolutely right. See the chapter on gravity where all will be revealed. The second effect, the appropriation of extra Zerons into the nucleus leading to an increase of mass is governed by Einstein's $E = mc^2$ equation. We will shortly be examining the mechanics of this mass again in more detail.

Newton's third law namely that any force has an equal and opposite reaction is an absolute law that is valid for any material system including a Zeron system.

Light at the end of the field.

One of the most valuable concepts of conventional physics is that of "fields". The puzzle is that fields seem to have no meaning or real existence other than an effect that could be detected with some sort of instrument. A well known but tongue in cheek definition is that a field is "the recipient of a set of equations". One of the more satisfying outcomes of Zeron theory is that it presents a lucid explanation for fields.

Let's start by an examination of "empty" space. If we could find a volume of empty space away from any gravitational influence and not touched by light or any form of radiation, we would find Zerons in constant movement, rebounding off each other in a perfectly random but resonant manner. Though we can get closer to these conditions out in space than on earth, a pattern of perfectly random Zeron impact is certainly nowhere to be found in our universe. As soon a there is an impingement of gravity, a ray of light a cosmic ray or a magnetic influence into our space, a pattern is imposed on the randomness of the Cosma's resonance. This non-random pattern of Zeron momentum is what we recognise as a field. This leads to a definition

A field is any non-random pattern of Zeron momentum transfer.

While the different types of field are determined by different patterns they all have at their root this Zeron impact signature. We are familiar with electrostatic fields, electromagnetic fields, and gravitational fields. These are all manifestations of a pattern within the otherwise random Cosma

One new type of field needs to be introduced. We have called this the quantum field. We saw that the presence of an atomic nucleus sets up a resonant Zeron shell. Similarly the presence of *any* body having mass sets up a disturbance in the "randomness" of the Cosma. This effect is present both at micro and macro scale. Because it originates at the micro level and therefore is subject to quantum effects we have named this disturbance the "quantum field".

Like gravity, the quantum field reduces in strength as the square of the distance from the object. However in other respects the quantum field is quite different. The basis for the quantum field is the same as that for the electron shell. We know that when an atom is excited by an input of energy the electron may jump to a larger "allowable" orbit. For the hydrogen atom the $n=1$, $n=2$, $n=3$, $n=4$, and $n=5$ orbits are distanced from the nucleus as the nth

power of 2 i.e. in the proportions 1,4,8,16,etc. Zeron theory proposes that in addition to these major high-energy Zeron shells other "ghost" shells are in place, having progressively larger radii. These shells have insufficient energy to permit the formation of an electron. However they must exist. This follows from the fact that we are, with the Cosma, dealing with a loss-free system. To assume that the resonant shells simply end at the last electron shell would be completely untenable.

If the radii of these shells increase as the nth power of 2 it will be seen that the radius soon leaves the realm of the atomic scale and will extend appreciable distances into the space surrounding the nucleus. For a massive body these quantum shells combine to form an enlarged "ghost" image of the object reaching far beyond the confines that we can see or detect by any but the most sensitive apparatus.

The quantum field of a massive body forms a "mantle" from the summated resonant shells. Being formed in a lossless environment, this mantle theoretically extends to infinity. However practically, the mantle can be considered to extend to that point where it is no longer capable of producing quantum effects. This system of Zeron shells constitutes the quantum field.

The quantum field is of great importance in considerations of "indeterminacy". It will be the main tool for the return of the universe to an objectively real status. In the interim there is an interesting perspective of what an atom looks like at close quarters and what the same atom looks like taking the extended quantum field into account. Diagram (a) is the atom at close quarters showing the nucleus and electron shells at radii of 2, 4, 8, units etc. Diagram (b) incorporates diagram (a) at it centre, but the quantum shells identified are at radii of 256, 512, 1028 etc. The diagrams appear to be identical.

DIAGRAM 12

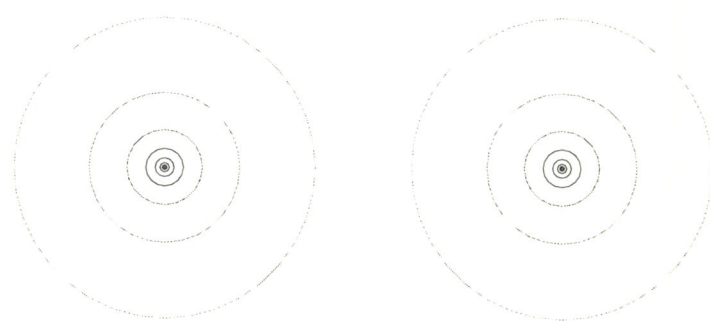

The quantum mantle is one of the stranger outcomes of Zeron Theory! I can foretell that those who make their living by promoting the paranormal will pounce upon this characteristic. There is nothing strange or abnormal about what might be called a "halo" or "aura"! It's the Quantum Field. However the possibility of detection by human or superhuman senses is extremely remote. These effects cannot be detected except in experiments sensitive enough to measure the smallest possible quantum effects.

Making light of the atom

Conventional physics has it that when an electron changes orbit from a higher state of excitation to a lower state, a photon is emitted. Conversely, a photon can cause the electron to jump to the next possible orbit. The whole concept of light emission from the atom in conventional physics is problematic in the extreme. Firstly when an atom emits light, CP (conventional physics) maintains that it emits it equally in all directions. At the same time, CP states that if you measure the light emitted at any particular spot you detect a "photon" of light that behaves like a particle. If you detect at a different spot, you still detect a photon. Does this mean that there are an infinite number of particles being emitted in all directions? Even more intriguing, how does the scientist ensure that the photon from his (omni-directional) light source goes where he wants it to? Might the photon not hare off in any one of the infinite alternative directions? No answers are to be found in CP textbooks.

By contrast, Zeron theory provides a clear representation of the physical processes involved and explains why the photon in the physicist's experiment will appear exactly where he wants it is to be! Sounds like mind control. Quantum physicists would be delighted by the incorporation of the experimenters mind into the experiment. However there is a perfectly logical and simple answer that does not involve the supernatural.

DIAGRAM 13

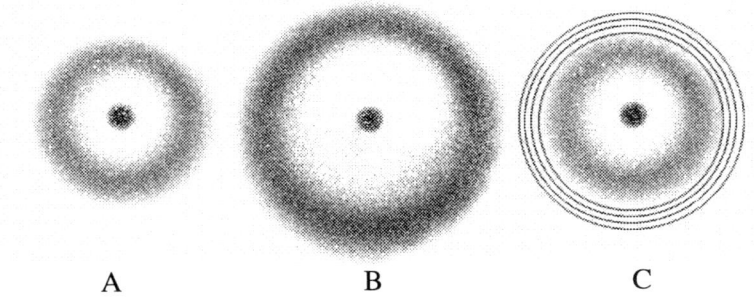

A B C

Emission of Light from an Atom In a stable situation we have three resonating systems associated with the atom. There is a resonating Cosma, a resonating nucleus and a resonating electron shell or shells (A). As soon as a stable atom is immersed in a higher energy environment (increased Zeron density in the surrounding Cosma), the resonant Zeron "electron" shells and nucleus find themselves out of step with the frequency of the surrounding Cosma. This is because increased energy means higher Zeron population density around the atom and therefore higher average frequency of fast-Zeron resonance in the Cosma.

The resonant Zeron electron shell responds. As it is no longer in synchronization with the surrounding Cosma it absorbs Zerons until the next highest resonant energy level compatible with the frequency of the surrounding Cosma is reached. The Zeron shell jumps suddenly to the next larger "permissible" radius (B).

It will be remembered that the resonance in the electron shell consists of tangential vibrations. Its new radius is determined by the next whole number of tangential standing waves in the shell.

Energy in whatever form consists of "packets" with a discrete energy and length of emission. The wave packet now passes the atom by, the energy in the surrounding Cosma drops and the electron shell once more is out of synch with the surrounding Cosma. It has to shed Zerons and shrink to its original radius. It does so by emitting a series of waves into the Cosma. These waves are transverse in character and continue to be emitted until the electron shell has finished shrinking. The nucleus and the shell once more become resonance-matched with the surrounding Cosma and the process is complete. The collapse Zeron electron shell to the lower energy level (C) creates:

An outwardly traveling Zeron impact waveform consisting of 10^5 to 10^6 tangential oscillations in the form of an expanding shell.. We recognise this as light.

This expanding shell is the "wave packet" of conventional physics. Once the electron Zeron shell and the nucleus are in synch, no further energy output takes place. The start and end of the emission are stable resonant states with zero emission.

What about the problem of the obliging photon that moves precisely to where the physicist wants it to? Well, no matter where he places his photon detector, it will be impacted by the expanding omni-directional light shell. Thus it seems very likely that what the experimenter is detecting this wave train and not a discrete photon particle. Confirmation of this is to be found in the next section

Isn't this all just conjecture? Well all models are, to greater or lesser degree, conjecture, but this one makes such good sense! However one of the real strengths of Zeron modeling is that from time to time we can positively tie these models back to the results of conventional experiments (even if these results are not fully understood by the scientific fraternity that produces them).

If we look at the Zeron based process of excitation, emission of light and a return to a stable state in more detail, one would expect to find an initial build-up of wave energy as the Zeron shell initiates its collapse followed by a period of steady emission as the shell steadily reduces in radius over a finite time, followed by a reducing wave amplitude as the electron shell gets back into total synch. This is exactly the pattern found in a photon wave packet as determined by experiment. For my money, this lends considerable veracity to what otherwise might appear to be an enormous thumb-suck.

DIAGRAM 14

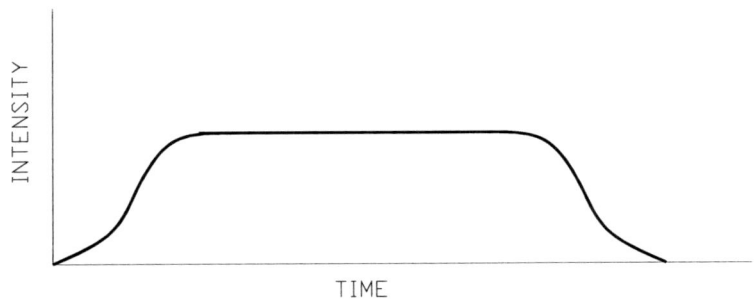

PHOTON INTENSITY WAVE PACKET

Further evidence is provided by the well-known fact that light waves are transverse in nature. The outwardly expanding (light) energy waves are a reflection of the action of resonance patterns within the electron shell. The resonance within the shell is essentially contained within the narrow band constituting the shell. The shell has very limited radial dimension, and is therefore essentially tangential / orbital in nature. Thus the waveform of light that has its roots in the resonance within this shell is also essentially transverse to the direction of travel. This agrees well with conventional knowledge.

Here for the first time we have a visualisable model of the processes involved in the emission of light from an atom. Heisenberg was correct when he said that the model of light emission from an atom was sheer madness. The Zeron model has returned us to sanity

Waving particles

Light is known to manifest as an energy wave and also in some circumstances as a particle. This puzzling phenomenon of wave/particle duality not only appears to be valid for electromagnetic radiation but is also thought to be applicable to all particulate matter. Young's double slit experiment producing interference bands from a beam of light is the classical proof of light's wave nature. Einstein's photoelectric theory (he was awarded the Nobel Prize for this discovery and not for Relativity) backed up by experiments by Millikan showed unequivocally that light packets or quanta (photons) behave as particles. Who was right? The particulate nature of light was confirmed by Compton who showed experimentally that a Photon and an Electron in collision behave as two particles, as conservation of energy and momentum was shown to apply to the incident. How are the results of these independently conducted but conflicting experiments to be reconciled? Zeron theory reconciles these in a rather unusual way.

We have seen from the above Zeron theory of light that a resonant expanding Zeron shell is emitted from the Zeron electron shell as it collapses to a lower energy level. The resonant spherical outwardly expanding light shell does not transport Zerons. The Zerons within the Cosma only serve to pass on the Zeron momentum by means of Zeron-on-Zeron impact. Apart from their normal resonant movement, the Zerons in the Cosma remain on average, sensibly where they are as the spherical light wave passes though. However, if the positions of individual Zerons were to be monitored, it would be found that if the normal vibrational movement is averaged out or ignored, all the affected Zerons will travel on average one "pitch" (the average distance between two Zerons in the Cosma.) per oscillation of the light wave.

At the end of the line, when the wave-packet hits a particle detector or another particle, the foremost Zeron in line dumps its kinetic energy and so does each of the subsequent Zerons as they travel one pitch and collide with the body to which they are aimed. The aggregate of the transfer of momentum for the entire wave train is what is recognized as the energy of a photon and is detected as the dynamic impact of photon particles. The rapid "machine gun" impact of Zerons is fast enough to be, in effect, an instantaneous event in the life of a comparatively much more massive electron or other particle. The particle behaves as though it has been hit by a particle whose mass is equivalent to the sum of the Zerons that have impacted on it.

The surprising conclusion from this model is that that there is actually no such thing as a photon! For the present, however, we will equate the machine-gun impact signature of wave train Zerons to the photon of conventional physics. To distinguish our model of the "photon" from the conventional photon, we have called ours the v-photon (virtual photon). The question might well be asked, *when* does this v-photon form? As in the case of the electron,

The v-photon will form in any experiment designed to detect photon particles!

It is the experimental apparatus that causes the photon to be detected as a particle. Only a piece of apparatus built to detect photons will detect these as light "particles". This is equivalent to determining the position of a photon. All other light detecting equipment will measure the movement of a traveling waveform and the particle will not be perceived to coalesce. This is equivalent to measuring the "momentum of a photon".

Here we get to the heart of the uncertainty principle. The Heisenberg uncertainty principle states that the photon (considered at all times to be a discrete particle) appears not to possess momentum and position at the same time! How on earth does a particle get to drop one of its fundamental properties? Conventional physics is silent on the matter. It just happens but no one is saying *how* it happens. Here at last is the solution to this rather strange statement. It is not that the photon does not possess momentum and position at the same time, but that the photon particle (whose position can be determined) and the wave-train (whose momentum can be determined) simply do not exist concurrently! It's one or the other. The wave train has momentum. At the end of the line when the "particle" is detected, it has position.

Is this splitting hairs? Certainly not! Uncertainty theory lends some sort of mysterious and unexplainable characteristics to the photon. All is indeterminate and immeasurable and by extravagant extrapolation by conventional physicists, this applies to the entire universe! By contrast, Zeron theory has provided a lucid and concrete explanation. Simply put:

The v-photon physically exists either as a waveform or a virtual particle but not both at the same time.

The above is a bit simplistic because wave trains emanating from any source cannot be beam-like. *Any* source whether composed of many atoms or a single atom emits the wave packet as an expanding spherical shell of light. This explains why the physicist can place his detector anywhere in relation to the source of light and be confident that he will detect a "photon"!

Chapter 15

The Puzzle of the Slit Machine

> *The Quantum explanation of the single-photon-at-a-time experiment is as fuzzy as quantum theory itself. We examine the impossibility of producing a single photon emission in any experiment. Zeron theory's electron shell solves the problem of single photon interaction through two slits. We compare the electron and photon "particles". Particle detectors will de facto detect particles and not waves. Analysis of a similar experiment, the polarized light also yields to Zeron theory. Why Alain Aspect's experiment proving indeterminacy is invalid.*

Zeron theory enables us to sweep away some of the apparent inexplicability of modern physics by re-interpreting certain key experiments. These experiments purport to show that matter at the atomic level has an inherent indeterminacy that prevents us from ever being able to describe with certainty, the physical phenomena that are part of our lives. The first of these is Young's famous double slit experiment and its refined versions.

To summarize the experiment briefly a ray of light is directed towards two narrow slits in a screen placed in front of a second screen or light sensitive plate. The light passing through the slits produces an interference pattern of dark and light bands due to the destructive interference and constructive reinforcement of the light waves, which travel slightly different distances on the different paths to the second screen. This experiment demonstrates the wave nature of light in a convincing fashion. If one of the slits is closed the pattern disappears.

The bizarre part of this experiment is that if the intensity of the light beam is reduced so that only one "photon" at a time hits the second screen and if both slits are open, the light dots (recorded sequentially and photographically on the second screen) summate to present an identical interference pattern! As soon as one slit is closed, the pattern of dots changes dramatically and the interference pattern is no longer to be found. Identical results are obtained if electrons are used instead of photons. This is a real puzzle. How can particles produce the same interference pattern if they go through the slits one at a time! With what are they interacting?

The result is interpreted to mean that when both slits are open the photon would need to have passed through both slits at the same time, clearly an untenable position. To explain this apparent anomaly a complex line of reasoning is adopted. The conventional explanation has its roots in quantum physics and is as follows:

In the quantum physics world it is impossible to measure the location and momentum of a particle (such as the electron or photon) at the same time without introducing in either measurement a degree of uncertainty. Any attempt to measure either quantity accurately results in the corresponding inability to measure the other quantity accurately. There is an element of uncertainty in the quantum world of these particles that no cleverly designed apparatus can circumvent. In 1927 Heisenberg quantified this uncertainty with his famous "Uncertainty Principle" Following on from this, the conclusion was then drawn that the electron / photon simply did not posses both a momentum and a position simultaneously. No satisfactory explanation of this puzzling statement has ever been made. This inability to determine both the momentum and the position of a particle has been interpreted to mean that the particle (photon or electron) leaving the light source in Young's twin-slit experiment has no fixed trajectory and therefore each particle possesses an infinity of different paths. The paths thread through both slits in the screen and "encode" information about each other resulting in the interference pattern. No theory is advanced as to how the encodation takes place, or how a sequence of particles emitted one after the other can possibly encode each other. The explanation is as fuzzy as the quantum description of the photon.

Zeron theory clearly models the mechanism behind this phenomenon. We have described the photon as being primarily a translocation of Zeron momentum in the form of an expanding spherical shell-like waveform. No photon particle exists during this phase of light emission. In a sense the Uncertainty Principle is absolutely accurate. One cannot determine the location of a particle that doesn't exist! The scientist generates light and then places the slits in front of the light source. For as long as there is plenty of light output there is no problem in explaining the diffraction pattern by (correctly) invoking the wave like properties

The Puzzle of the Slit Machine

of light. The problem comes with this single photon / electron emission.

Before we proceed to build on this particular model, the setup of this twin slit, low light level experiment needs critical appraisal. Firstly it seems incredible that the experimenter should assume that a point light source emitting photons, one at a time, would conveniently steer the photon "particle" towards the slits! Any point light source will emit in all directions even at the "one photon" level. How then does the photon land up at the slits? No matter what focusing or directing apparatus the physicist sets up there is no guarantee that the photon will not simply take off in another direction. With all the directions available in three dimensions the correctly aimed photon would be the exception and yet the photon faithfully lands up where intended every time. There *has* to be a better explanation.

The spherical Zeron wave progresses omni-directionally and will arrive at the slits no matter where they are placed in relation to the light source. Even at the minimum "one photon level" the waveform is fully spherical and fully developed. The "one photon" waveform passes through both slits creating exactly the same resonance patterns and conditions that create the interference pattern at higher light levels. There is no difference in the process between the multi- photon case and the single photon case. In both cases it is the constructive reinforcement and destructive interference of the Zeron shell passing through the slits that determines whether the photon will "coalesce" at all in any particular place. QED.

One of the problems that physicists would still have with this experiment is the one with electrons. The validity of the experiment is based on the premise that each electron is a particle. How is it then that in the single electron experiment the electrons coalesce at different places on the second screen? Here the problem arises because the incorrect premise that it is electron particles that are making their way one by one towards the slits The Zeron theory position is that the electron particle does not exist until it hits a particle detector.

Before proceeding further it might be useful to draw the distinction between a light wave and an electron wave. Light waves are essentially the propagation of Zeron kinetic energy by means of Zeron to Zeron impact. The Zerons do not sensibly move with the wave. By contrast, propagation of the electron wave is essentially a translation of Zerons through space. Whether it is Zeron momentum that propagates in waveform or translation of the Zerons themselves in waveform, the results on passing through the two slits are identical. Conventional physics assumption that it is electron particles that are sent through the slits is false. The electron *wave* passes through both slits producing the expected interference effects and the electron only forms downstream of the slits when the electron waves encounter the detection screen. This is in line with quantum theory that insists that the apparatus used will determine what is detected, downstream of the slits, waves or particles.

Imagine for the moment that you are designing apparatus to show the self-interference of electrons as proof of quantum theory. Your first assumption will have been that electrons are particles of matter. Your emitter will have been chosen as being capable of emitting electron particles. How do you know whether it is waves or particles that are being emitted by the apparatus? Well you have experimented with the apparatus before and surprise, surprise, particle detectors have shown that it emits electrons! So what sort of detector will you be placing downstream of the slits? Probably a Geiger counter or a scintillation detector or some such. These are designed to detect particles. Guess what you will detect? Of course if one was to measure waveforms from the emitter and place a wave detector in place of the particle detector it could be shown that what is passing through the slits is an electron waveform. To the best of my knowledge this has never been tried.

The final result of believing that particles are producing interference patterns is inevitable. The physicist believes he is dealing with particles. The particles behave like waves. Confusion reigns. Not to worry. Simply invoke the Uncertainty Principle, and then anything is possible! In sharp contrast, Zeron theory has provided a simple, lucid explanation of what is really going on.

There's more. Recapping on the Zeron theory of fields, it is certain that in a loss free system, resonance shells exist well beyond the immediate vicinity of the atom. We have named these "quantum" fields. The shells *have* to be there because of the loss free system of the Cosma. This has a direct bearing on Alain Aspect's experiment that purported to show that faster than light co-operation between systems was possible. Aspect set up two

The Puzzle of the Slit Machine

identical racetracks for photons, replete with resonating optical switches so arranged that that they ensured that the photons could not "communicate" at light speed. If the photons were to communicate they would have to send a signal faster than the speed of light over the 15 meters separating the photon detectors at the end of the track. It was found statistically that the photons had inter-reacted. At this juncture the universe receded into indeterminacy!

One of the assertions of quantum physics that we really identify with is that the apparatus is an integral part of the phenomenon we are trying to examine. Aspects apparatus is unlikely to have produced valid results. We say this for two reasons.

1. Resonating mirrors were used as an optical switching device. If any arrangement would ensure the co-operation between the two arms of the experiment, this was it. Resonating *anythings* set up far reaching influences in a loss free system.

2. Aspect did not know about quantum fields. Fifteen meters would be well within reach of a powerful enough quantum field. It is instructive to note that at very low light levels, Aspect was unable to get any correlation between photons. He simply increased the level until the experiment worked! In doing so he simply increased the strength of the quantum field until it showed the desired result. Had the experiment been valid, it should have worked at the lowest light level.

Clearly at the lowest level, the quantum field of both the apparatus and the particle did not have enough energy to inter-react and influence the result. At higher levels they did. The two arms of the experiment were enveloped in a standing wave configuration of the Quantum field. In a standing wave configuration communication is instantaneous because no signal has to travel. This concept might be easier to understand in the context of the way gravity works. The reader is encouraged to read the chapter on Gravity for an explanation. The photons did not have to communicate to co-operate. They were immersed in identical standing wave configurations and behaved accordingly. Ironically Aspect's experiment, if it showed anything, showed proof of the existence of the quantum field! Objective reality has once again prevailed and the universe is safely returned to determinacy!

Another Achilles heel for the conventional physicist is one in which light is passed through two polarizing filters. Polarizers have the capacity to take a light wave and filter it into waves that operate in one plane only. In the experiment, polarized light is passed through two polarizing screens. If the planes of polarization of the screens are at right angles to each other, no light is passed. If the planes are parallel, all the light is passed. At a relative inclination of 45 degrees between the polarizing planes of the screens, half the "photons" pass through. The question is which half?

No use saying to the scientist "does it matter, half is half!". This very simple experiment creates huge difficulties in conventional physics. The photons are all supposedly identical. What decides which photons will pass and which will be obstructed? If one polarized photon at a time is directed towards a second polarizer and is either passed or blocked, argument is that we cannot describe the photon other than in terms of a probability as to whether it will pass or be blocked. Therefore the photon, and by (gross) extrapolation, matter in general, does not have determinable properties. Once again the Heisenberg Uncertainty Principle is invoked and our universe recedes into indeterminacy! From small things come big results!

Zeron theory provides a very simple explanation. Sorry, chaps, there is no such thing as a photon particle. What you perceive as being a particle is a light wave of very short duration that contains within its makeup the "one pitch per wave" series of Zeron impacts that you interpret as a particle. It remains a light wave. All that happens when the polarizing screens are at 45 degrees is that the amplitude of the light wave is reduced and the transmitted light is perceived to be half blocked. I'm afraid that our friend Heisenberg has not got an uncertainty to stand on. Those aspects of quantum theory that rely on uncertainty must be rejected. Consequently there can be no universal extrapolation of the Uncertainty Principle and saying that indeterminacy is an inherent property of matter can be shown to be invalid in whatever guise it may appear. The universe can be considered to have returned to a rational and objectively real status!

Chapter 16

A New Relativity

We re-examine the Michaelson and Morley experiment in the light of Einstein's Special Relativity equations and the Fitzgerald-Lorenz contraction equations. Einstein didn't need an Aether for his relativity theory but never denied its existence. Re-examination and refutation of Einstein's spaceship thought experiment. The special status accorded to the observer is a weak point of Einsteinian theory. Relativity is a good calculating tool but does not represent reality. All boils down to time being variable or invariable. We re-define time as an invariable. Establishment of Zeesec as a unit of time. We establish the practical possibility of establishing a value for the Zeesec as well as the possibility of establishing the existence of the Cosma. Treating time as a variable always leads to a lack of objective reality.

Here we tread on holy ground indeed! The theory of relativity has captured the imagination of the world. Einstein turned physics on its head and his audacious theories were generally accepted in the scientific world, as they are today. One has to wonder how it was that a theory, so profoundly counter-intuitive came to receive such wide acceptance. Probably the stature of Einstein had something to do with it because prior to the publication of his special theory of relativity, Einstein had been the recipient of a Nobel Prize for physics. Perhaps it was the very counter-intuivity of the theory that made it so hard to challenge. Did anybody really understand it? Would any scientist who in his heart of heart did not understand it or doubted its veracity, put his reputation on the line by opposing the theory? It was another case of the Kings clothes. Fortunately for those of us who are not physicists, there is no reputation to lose. However, though it is not without some trepidation that we challenge the theory, the results make such good sense that the effort is eminently worthwhile.

You will remember that Michaelson and Morley raced two light beams at right angles to each other to try and find some effect on the speed of light due to a supposed Aether. They found none.

When the Michaelson and Morley experiment was performed there was much speculation as to what the results meant. Four possible explanations were given to explain the results in a way that would enable scientists to retain the concept of an Aether.

The first was the proposition that the earth was fixed in the Aether and that everything else moved with respect to the earth. The M&M experiment would then be unable to detect a difference even if there was an Aether. This was rejected on philosophical rather than on scientific grounds, proving that even physicists become philosophers when pressed hard enough. The philosophy was that it would be difficult to concede that the earth, one of several planets spinning around a very minor star in a vast galaxy of stars could have such a unique status. In effect acceptance of this proposition would place the earth at centre of the universe. This was simply unacceptable.

The second proposition was that the Aether was dragged along with the earth, making the Aether flow unobservable. This was rejected on the technical basis, which we won't go into here, that if this were so the well-known phenomena of the apparent seasonal change of position of stars (aberration) would not be observed.

The third proposition was that the speed of light was constant with respect to the source. This was rejected because this would require that the speed of light could assume different values through the Aether depending on the velocity of the source relative to the Aether. The results of the observation of double stars finally discredited this theory. Certain double stars are twinned and they rotate rapidly so that one star is retreating from us as the other is approaching. The light received from a twin star has a characteristically variable intensity that assumes a perfect sine wave profile. This would not be the case if the speed of light were dependent on the speed of the source. The sine wave would be distorted by any variation in the speed of light caused by the approach of one star and the simultaneous retreat of its twin.

The fourth and ultimate solution was provided by the Fitzgerald-Lorentz contraction equations of which we have previously made mention . This provided a solution by saying that there was a contraction in the length of the interferometer arms of the Michaelson-Morley experiment when they were orientated in the direction of the Aether flow. This foreshortening would exactly compensate for the effect of a moving Aether on the velocity of light.

Albert Einstein by his Special Theory of Relativity (STR) confirmed that this contraction would indeed take place, but at the same time, by sleight of brain, he dispensed with the necessity of having an Aether at all! In consequence, the conventional physicist of today believes that there is no Aether. Ask the physicist why he believes that to be true and he will probably tell you that Einstein proved that there is no Aether. In actual fact Einstein did not prove that there wasn't an Aether. He merely stated that *for his purposes*, it didn't matter whether there was an Aether or not.

Before proceeding with an examination of relativity in the light of Zeron theory, it is perhaps useful to examine some of the paradoxes that can arise from Einsteinian relativity. Much of Einstein's work centers around the "observer" and what is observed. What is observed, (i.e. observations carried by a light signal) is then equated with what is actually happening. As with much of modern physics we have to mix philosophy and science in order to bestow credibility on a theory. The philosophy part is that the scientist "can only believe what his instruments tell him." This means that the observation is believed and the belief is equated to reality. That this is a shaky foundation can be illustrated by the unassailable fact that things are rarely what they are observed to be!

One consequence of this semi-philosophical approach was that *time* came to be considered as a variable. This leads to various hard-to-comprehend and sometimes apparently illogical conclusions.

DIAGRAM 15

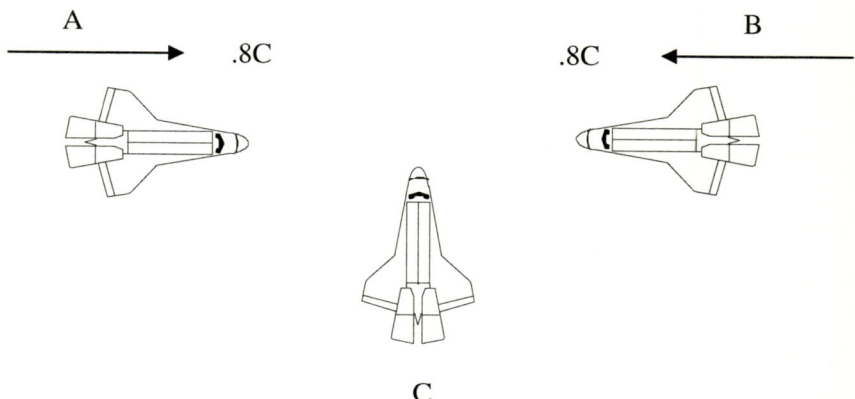

Closing Velocity of Two Spaceships

As an example, consider two spaceships A and B approaching each other with each traveling at 0.8c. The closing velocity is surely .8c+.8c? According to Einstein an observer on ship A, will measure the approach velocity of ship B at 0.98c. This is profoundly counterintuitive and is a direct result of treating time as a variable.

The fallacy of this result can be shown as follows. Say we place a third spaceship C midway between A and B and station an observer on the central spaceship C. We now have a closing velocity of .8c between C and either of the spaceships A or B. It will be impossible for an observer on C establish by looking at spaceship A whether he is moving towards spaceship A or whether A is moving towards him. All he has is an apparent relative closing velocity between him and the other spaceship. *We* know (as the persons setting up this thought experiment) that the closing velocity between A and C is 0.8c. If we now compute the speed of either A or B as measured by the observer on C we will find that according to Relativity Theory, spaceship A is approaching the observer on spaceship C at an apparent velocity of .689c. Likewise, spaceship B is also observed as approaching C at a velocity of .689c. According to Einsteinian thought, these closing velocities constitute the "reality" of the situation, as an observer can only believe what the (light activated) instruments tell him. The reality that C has established is that the closing velocity of the two spaceships A and B is now .689c + .689c = 1.378c since he has looked at both and found that each is approaching him at this velocity of .689c spaceship C decides that he is in considerable danger and he moves off quickly. Relativity Theory tells us that a closing velocity of 1.378c between A and B is impossible. We'll have to start again since we know now that A and B each have a velocity of .689c (C told us and he's a very reliable observer). If A now looks out the window and sees B approaching, he will establish that B is approaching him at a velocity of .934c. What? I thought the closing velocity was .975c! And yet, nothing has changed. Simply by inserting and removing the central spaceship with its observer, we can compute a whole series of velocities for the outer spaceships - clearly an impossible result. Einstein would probably point out that if you remove the observer C, *his* truth is no longer valid. The only conclusion if this is the case is that Einstein's truth is relative and is no longer the truth. Truth *has* to be absolute if it is to be true. Relative truth is a lie.

Another simple example serves to show the anomalies of relativity. It concerns the increase of mass with velocity. We are told that the universe is expanding. The further away the remote galaxy from us, the faster it recedes from us. This is measured by the "red shift" in the light reaching us. In general the velocity with which heavenly bodies recede from us is Hubble's constant times the shift in frequency of the spectral lines towards the red end of the spectrum.

Say a relatively nearby galaxy recedes from us at 1/4 c. From Relativity Theory we can compute that the galaxy has an apparent mass that is increased by approximately 15% because of the speed of the galaxy relative to us. Conversely our galaxy and everything in it would, to an observer on that galaxy, have a mass 15% greater than the mass we observe for ourselves. This, according to Einstein is "reality" because we and "they" have to believe their light driven instruments.

If we now look at a further galaxy the red shift indicates a relative velocity of 3/4c. Their galaxy is observed by us and our galaxy is observed by them to have a mass increase of 50%. Now clearly we cannot simultaneously have a mass that is increased by

15% and 50% at the same time. The mass increase is purely perceptual and not actual.

Equating perceptions to reality is like saying that our ocular systems and the neuron networks of the brain that interpret images (our perceptions) determine the manner in which the universe behaves. Were the whole of mankind to disappear tomorrow, the universe would undoubtedly continue to exist without our precious perceptions, or us.

The anomalies above are caused by changing the frame of reference. A frame of reference relates to everything in the vicinity of an observer that is observed by him. It is as if, for the purposes of calculation, the observer becomes the centre of the universe. Einstein by equating observations to reality, has in effect, promoted the observer to this elevated position. This is exactly equivalent to making our observer the centre of the universe. It's untenable. If these anomalies, caused by changing frames of reference are to be removed, a universal frame of reference needs to be established - a seemingly impossible task. What in the universe could we possibly perceive that provides the same base for measurement for all the stars and galaxies? Everything is in relative movement and subject to all the aberrations that such a movement causes. However we shall see that provided we are prepared (with good grounds) to consider time to be a non-variable, the concept of a universal frame of reference, however unlikely that may seem, is not an impossibility!

Before we re-visit relativity in detail, it is necessary to embark on a semi-philosophical argument regarding perceptions and time. In Einstein's Relativity Theory and indeed in many physics theories, there is a philosophy that says "If we can't measure it then it doesn't concern us as physicists" or "a phenomenon is not a phenomenon until it is a measured phenomenon". This is taken a step further in conventional physics by saying that if a phenomenon is of no concern to us or is not measurable then it might as well not exist. This is usually followed by "the phenomenon doesn't exist". Such was the fate of the Aether.

In Relativity Theory, it is similarly assumed that if an occurrence is perceived by the receipt of a light signal, then whatever perception is obtained represents the truth of what is actually happening. For example if two rockets travel at high speed relative to one another then the observations carried by light beams will show that rocket A will see rocket B as being shorter than itself and visa versa. Now, it is stated this is a real effect because there is no way of determining otherwise since we cannot get any signal faster than light! Read any article that purports to explain relativity. The arguments that lead up to the relavistic equations are peppered with words such as "appears to" or "sees" and the discussion inevitably flips back and forward between an observer within the experiment and the person presenting the thought experiment who does so from a privileged outside observation position. This is this sort of sleight of mind that leads to illogical results. In fact all of the Theory of Relativity is based on what "appears" to happen as determined by the passage of a light signal.

The strength of the Theory of Relativity is that it can be a magnificent tool leading to equations that describe physical phenomena most of which can be and have been tested in practice. But to say that the equations enable us to interpret what is really happening at the atomic or subatomic scale simply does not hold water. The theory remains a useful tool but tells us little of what is happening in an "objectively real" sense. In the end the whole question of the reality or otherwise of Relativity results boils down to our definition of time.

There is a fundamental choice to be made. Simply put, if we choose to make time a variable, we have no need for a fixed reference frame and the results of Relativity Theory represent the objective reality of what is happening. This is the Einsteinian view. If on the other hand time is not a variable but can be defined in absolute terms, then for the objectively real scenario, we have need of a standard reference frame, a seeming impossibility. No wonder Einstein chose variable time! In contrast, we have picked up the challenge of finding the universal reference frame.

Before we can proceed further, it is necessary to examine and define time. We hold that to consider time as a concrete entity to which we can attach a variable (non-constant) symbol "t", is simply incorrect. We can identify mass and allocate a symbol to it because exists physically. Time however has no physical properties at all. It lives in the minds of men as an abstract concept. There can be no justification for assigning symbol "t" to "time" thereafter proceeding to treat it as a mathematically manipulable entity.

The "objectively real" definition of time is as follows:

Time is the measure of the interval between two successive events.

Leading from this definition the unit of time can be defined as follows:

The unit of time is an empirical proportion of the interval between two events.

Time is not seconds or days or years or any other particular empirical unit. Time is essentially an interval, and the interval between two successive events cannot be changed however we might want to measure it. There is nothing we can do to alter the interval. One thing happens and then the second and if we try to change the interval between first and second then we are looking, by definition, at events 1 and 3 not 1 and 2. Any discrepancy there might be in the results of measurements we perform to measure the interval, must therefore be ascribed to the measuring instruments or to perceptions. It is with this concept of the invariability of time that we drag ourselves back from the morass of relativity to the realm of reality.

With this concept of the invariability of time, we undertake a major departure from conventional physics. "Einsteinian time" *is* a variable and it has been generally accepted as such by the scientific community. The term t for time is commonly used in physics equations and its value depends on the relative velocity of the observer. Our time (the unit of which we have named the Zeesec or t_z) is a non-variable. This might not seem to be of particular importance, but it is the key factor that, as we shall see, enables us to make real sense of relativity and other of the more mysterious propositions of modern physics. As trivial as this definition of time might seem, we shall see that it has the surprising but inevitable consequence that all other dimensions and many of the properties of matter generally considered to be fixed, now become variable! And Zeron theory was supposed to make all things easily understandable?

A New Relativity

As complicated as this may seem at first, we have here, the key to the unification of forces and the simplification of the many processes and phenomena in the physical world. Intangibles such as inertia drop their purely empirical/descriptive identities behind to take on concrete identities definable in terms of fundamentals and classical physics. Even Relativity loses its mystery to be explained in similarly classical terms.

In order for us not to leave the invariability of time in the realms of the theoretical, (this is a practical treatise) we need to establish that it would at least be practically possible to establish such an invariable time standard. As we shall see in the further development of Zeron theory, we cannot depend on any measuring process that depends on the shape size or properties of the atom. The properties of the atom vary according to the speed of the atom relative to the Cosma. Therefore the characteristics of any clock or measuring device made from atoms are also variable as will be the time measured by such devices.

In the entire universe the only reliable physical invariables usable for defining time is the speed of light and the speed of the Zeron upon which it depends. If we wish to establish time as an invariable it must be based on the speed of light or the speed of a Zeron. We can then define the unit of time as a fixed proportion (constant) of the interval taken for light to travel a fixed distance (constant) between two points.

We are now going to go into a bit of detail to establish whether a practical unit for non-variable time can indeed be established. This detail is not critical to what follows but it would be comforting to know that what we are talking about is not pie in the sky. Let us try and find a fixed distance over which to race a light beam and so establish a unit of time. For example if the distance is taken as the distance from the earth to the moon and back, and the proportion or constant is taken as 1/2, then the unit of time (called a Zeron-second or Zeesec) is the distance divided by the speed of light times 1/2. Taking the speed of light as 299800km /sec and the distance to the moon and back as 2 x 384400 km, the time for the return trip of a laser beam is 2.56438 seconds.

If we make the constant or proportion to be 1/2.56438 the Zeesec would be nominally equivalent to the currently accepted second of time, with an accuracy only dependent on our ability to

The God Particle

measure the distance to the surface of the moon. The ability to do this has recently been enhanced by the placing of a laser reflector on the moon's surface. How do we know that our instruments enable us to consistently measure the distance to the moon if the instruments themselves are subject to relavistic or other effects? In practice it does not matter how we measure the distance as long as we always measure the *same* distance exactly the *same* way. It is not the actual distance that matters but the consistency (constancy) of the parameters affecting the measurement of that distance.

Apart from the given of ensuring that the measuring instruments were always operated under consistent conditions, establishment of a Zeesec would involve taking into account any variations of the distance between earth and moon due to perturbations in the moon's orbit. One would always have to make the measurement when the target point on the moon was at the same orientation to the measuring instrument etc.

The situation is complicated by the fact that our companion in the sky does some very strange things. Apart from having an elliptical orbit, the moon is subject to a few major perturbations and some 1500 known minor perturbations. The question then arises as to whether we can expect to use the earth moon distance as a baseline. Laser measurements have enabled scientists to say with a high degree of confidence that the Earth/moon distance increases by about 30 cm per year. Now 30 cms compared to the earth moon distance represents about one part in 10p10. If the accuracy of that measurement is factored in, we should be able to establish a time standard to about one part in 10p12 which is about the same as the accuracy of the time standard established by a cesium atomic clock. Of course if we are not stationary relative to the Cosma we would have to make sure that the axis of the elliptical orbit of the moon points in the same *celestial* direction each time we take measurements!

This points to a potential way of analysing whether in fact we are moving through a Cosma. If the comparison between atomic clocks and the Zeesec shows variance with a period corresponding to the combination of the precessional period of the moons orbit and the earth's movement around the sun, we will have found general proof of the affect on atoms of our movement through the Cosma and, in fact, proof of the existence of the Cosma itself. If that sound too complicated, don't worry. We are getting somewhat overly technical here, but this is really intended as a bit of a challenge to the conventional scientific fraternity. But it *is* an intriguing thought that we now may be able to establish the presence of the Cosma (`nee Aether) by looking at the *apparent* perturbations of the moon. Fascinating indeed!

Back to the question of time. By defining time as a function of the speed of light and a nominal but constant distance, we develop a definition of time that is importantly distinct from that of Einstein's relavistic time. Our Zeesec is derived from the speed of light (constant), a fixed distance (constant) and a proportion (constant). The Zeesec is therefore fixed and is a constant.

Time as measured in Zeesecs no longer depends on relavistic considerations or the time measuring instruments in the frame of reference of the observer. Time is absolute, is no longer the fourth dimension, and no longer has a place in the equations that were used to derive the special Theory of Relativity. Three dimensions will suffice.

Einstein's Relativity Theory is surely one of those best established by many diverse experiments and practical applications. Like the curate's egg it is (exceedingly) good in parts. With one exception, (that of contraction of length) it is not with the theory that there is any argument but we believe there is a good case for re-interpreting it in terms of objective reality with the assistance of Zeron theory. The chief objection to the Theory of Relativity is the concept of time as a variable particularly if, as we see above, we can develop a non-variable definition of time.

Take the case of an atomic clock sent into orbit. Experiments have shown that the orbiting clock and a land-based clock do not agree, and that the difference is very close to the value calculated by the relevant Einstein's relativity equation. The question is this. If the apparent frequency of vibration of an atom, as determined by an atomic clock, has been found to change, is it the atom that has changed its behavior, or has time changed? We submit that it is the atom that has changed its frequency of vibration particularly as we have shown the possibility of a constant unit of time. Following on from previous paragraphs we also submit that time dilation as expressed by Relativity Theory is, in its essence, a meaningless concept, whereas the change in the natural frequencies of atoms would be a practical visualisable and objectively real description of that same phenomenon.

Treating time as a variable always leads to a lack of objective reality. Shame on you Einstein! Relativity will therefore be developed in a Zeron/Cosma environment with absolute time to see whether there are alternate descriptions that do not rest on variables such as Einsteinian time.

Chapter 17

Relatively Quantum

By considering the "foreshortening" equations at the quantum scale of an atom we find that there is a missing term in Einstein's relativity equations. The term takes into account the direction of observation and is expressed as Cos ϕ. We find that the atom has both foreshortening and after-lengthening as speed increases with a limiting length of 0.707 times original length at the speed of light. We examine the emission of light from an atom moving through the Cosma and show why the speed of light is constant relative to the Cosma. We show why an observer can receive light at greater than c but that he just can't measure this. We establish the Cosma as the universal frame of reference. We show how red and blue shift establish the closing velocity of spaceships traveling towards each other, not Einsteinian equations. We re-work Einstein's famous train experiment to show the difference between reality and perceptions.

Now things get really interesting because we have to challenge and modify Einstein's relativity equations! Don't stop reading just yet. We will present an alternative version that sorts out many of the intractable and counter-intuitive outcomes of Relativity Theory. It's worth waiting for.

As stated at the beginning of this book, Relativity and Quantum Theory are uneasy bedfellows. Zeron theory enables us to take relavistic effects right down to the quantum level and show compatibility between the two theories. We do this by examining the emission of light from a moving light source. The light source is a single atom that has been suitably stimulated. The analysis begins with Einstein's foreshortening equations that were derived (plagiarized?) from the Fitzgerald-Lorenz transformation equations. (It is interesting to note that Fitzgerald and Lorentz derived their equations on the assumption of the existence of an Aether!)

Proceeding with the examination of a moving light source, the first thing to be looked at is the contraction in length that occurs when a body moves through space. The foreshortening equation is probably the only one in the Einstein relativity set that has probably not been fully tested by experiment. One "proof" of the validity of the foreshortening equation concerns the dimensioning of high-energy atom smashers. It has been stated that ignoring the effect of contraction in length would have meant that the design of these high-energy atom smashers would have been faulty and they would not have worked. It is suggested however that by using of the formula for contraction in length what the scientists were really doing was taking into account the increase of mass at relavistic velocities. These two phenomena are equivalent in Einsteinian theory, the use of either in appropriate ways producing the same result.

This "foreshortening" effect is the one major aspect of Einsteinian relativity with which Zeron theory is in disagreement and it will be shown that the foreshortening effect is only a special aspect of more general phenomena of shape distortion of an atom or particle subjected to relavistic velocities.

We now have to break a basic rule for the writing of a "popular" scientific book. The rule states that there should be no equations because nobody will read them. This becomes a little difficult to do if the subject to be discussed is an equation! The reader will therefore have to bear with us while we take a look at some of Einstein's equations. If you don't fancy going through the equations, just read the text. You won't lose much.

The equation for contraction in length of a body traveling at a speed v relative to an observer is given by Einstein's equation

$$L' = L\sqrt{1 - v^2/c^2} \quad \ldots\ldots\ldots\ldots\ldots\ldots(1)$$

Where L' is the new length, L the original length, c is the speed of light and v is the velocity of the body whose length was L. The meaning of this equation is best explained by a few examples.

If the velocity of the body is very low then the term under the square root sign becomes 1 and L'=L. That is there is effectively no change in length. If we make V equal to half the speed of light, then L' = .86L. That is the body contracts by 14%. If we make v equal to .9 times the speed of light, the body contracts by 56%. For V= .999c the contraction is 96 %. It can be seen that as the velocity gets very close to the speed of light, the contraction rapidly tends towards 100%. What happens at 100% contraction? Nobody is saying because our Einstein has worked out that it would take an infinite amount of energy to accelerate the body to light-speed. As this is a physical impossibility, we needn't consider it! The question is not answered and it, like the Aether, is simply sidestepped. Einstein would have made a good rugby player! In fact theoretically, the length of the body becomes zero, a profoundly counter-intuitive result.

It appears from the derivation of equation 1 that this shortening effect is as seen from the "front window" of the observing body and has thus been correctly labeled by some as a "foreshortening" effect. It is what is apparently happening to a body as we approach it at high speed and we are observing through the "front window". What is not addressed is what the "rear window" apparent effect is. Clearly at right angles to the relative movement there is no relative speed *in the direction of observation.* The lateral dimension of a passing spaceship is unaltered. What happens at a 45 degrees or at a 180 degree direction of observation? If the equation is correct the 180 degree view is identical to the zero degree view. This is because if we make the v in the above equation negative, the squaring of it makes it positive again and the equation remains unchanged. This 180 degree "rear window" view is usually not addressed probably because of the perceptual difficulties involved in the macro system of "observing" a body by looking in the opposite direction! Mathematically of course there is no problem.

While we are talking here of macro effects, one of the promises of Zeron theory is the melding of macro and micro systems. Let us therefore apply the macro "foreshortening" effect to the micro atom and its electron shell.

If we were to place ourselves at the centre of a hydrogen atom and were static relative to the surrounding Cosma, we would observe the Zeron resonance shells equidistant from us in all directions. Should the atom get adequately stimulated due to an increased energy level in the surrounding Cosma, the shell would be observed to jump to the next stable radius, but would remain equidistant from us in any direction albeit at the larger radius. Let us now assume that the atom is accelerated for a limited time so that it moves through the Cosma at a constant velocity v.

Let us assume for the moment that Einstein's equation correctly represents the foreshortening of a macro body for the "front window" observation. For the body to be foreshortened, each of the atoms of which it is comprised is similarly foreshortened. So if we look in the direction of motion, we will see that the electron shell is now a tad closer to us than before acceleration. If we now turn 90 degrees to the direction of motion, there is no forward motion relative to the Cosma so the electron shell is the same distance away from us as it was before acceleration. There has been no bunching of Zerons in this direction. Now we have a problem. Einstein's equation simply does not address this case. We need to modify it to take the direction of observation into effect. Equation (1) is therefore simply a special case of a more general equation that can be formulated to take into account the direction of observation namely

$$L' = L\sqrt{1-(v^2/c^2)} \cos\Phi \quad \ldots\ldots\ldots\ldots (2)$$

where Φ is the angle between the direction of travel and the line of observation. In fact this $\cos\Phi$ factors in the forward component of the velocity v into the equation.

For $\Phi = 0$ (front window view) $\cos\Phi = 1$ and equation (2) becomes the same as equation (1). Einstein's equation agrees with the Zeron equation. No surprises here.

For $\Phi = 90$ (right angle view) $\cos\Phi = 0$ and $L' = L$. There is no foreshortening effect. By contrast, Einstein's equation yields the same foreshortening as for the front window view. Not a very logical outcome.

But for the modified equation, so far so good. The $\cos\Phi$ produces a logically correct result. *The real surprise comes from the rear window view.*

For $\Phi = 180$ (rear window view) $\cos\Phi = -1$ and the equation becomes

$$L' = L\sqrt{1+(v^2/c^2)} \quad \ldots\ldots\ldots\ldots (3)$$

This equation looks almost the same as equation 1, but there is a critical difference. There is a plus between the 1 and the bracketed terms instead of a minus. The effect of this is found when we make v=c, i.e. the velocity of the atom relative to the Cosma is equal to the speed of light. If we look at the rear window view, when $v = c$,

L' = $\sqrt{2}$L in the "rearward" direction!

Here is a radical departure from Einsteinian physics. The atom has zero forward dimension, but has a 1.414L (root 2 times L) backward dimension. Adding the forward and rearward dimensions, we find that the atom does not have zero length, but on the contrary, the length has changed to .707 times its original length! Now that sounds more likely.

These foreshortening, after-lenghtening Zeron-based equations govern the limits of the distortion of an atom traveling relative to the Cosma. This effect is active on both the Zeron electron shells and on the nucleus, which is the source of the Zeron shell resonance. Movement through the Cosma creates a distortion in the Zeron electron shell (and in its various higher energy orbits)

In general:

A foreshortening effect at the front of an atom moving through the Cosma is accompanied by an after-lengthening effect at the rear.

To recap, here we have one of the more significant disagreements between Zeron theory and conventional physics. According to the above derivation, as a body approaches the speed of light, the length of the body changes to a limit of 0 +$\sqrt{2}$ L divided by 2 = 0.707L and not to the counter-intuitive "zero" length as given by Relativity Theory.

So what happens to the increase of mass at relavistic speeds? Unlike Einsteinian theory Zeron theory shows that increase of mass is a separate and almost unconnected phenomenon! There's a detailed section ahead that describes the mechanism.

The God Particle

Moving lights – Framed!

Consideration of the mechanism of light emission from a moving light source leads to an explanation as to why light has the intrinsic value of c. It also leads to the apparently impossible, the Universal Frame of Reference! This means in effect that the playing field is level for anybody anywhere.

In previous sections we developed the Zeron based model of the atom and the mechanism for the emission of light. To summarize briefly, the nucleus of the atom is portrayed as an agglomeration of Zerons held together by Zeron pressure and resonating in sympathy with the resonant Cosma around it. It is enveloped by the Zeron electron shell or shells. A shell owes its existence to the interaction between the waves emanating from the vibrating nucleus and the Zeron oscillations in the Cosma. Light emission is caused by the raising of the energy of the Zeron electron shell to the next highest allowable energy level followed by a return to the lower energy level. The collapse to the lower level results in the emission of an expanding shell of light, which transmits energy by means of Zeron momentum.

When the atom is stationary relative to the surrounding Cosma, the nucleus of the atom is located at the geometric centre of a spherical or spheroidal Zeron electron shell or shells. Light is ejected from the atom at velocity c.

DIAGRAM 16

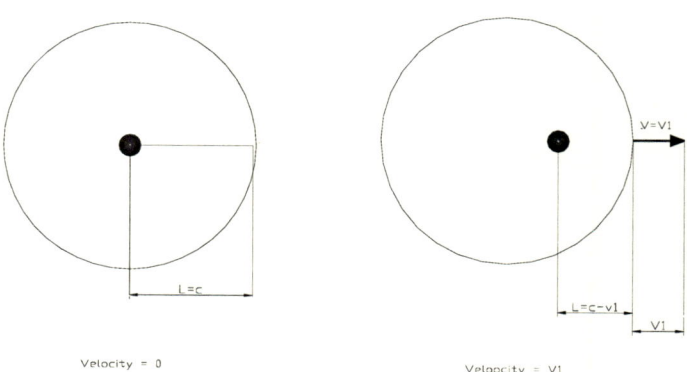

If the atom accelerates relative to the Cosma there is an immediate increase of Zeron "pressure" on the nucleus in front and a decrease of Zeron pressure behind. This distorts the nucleus and the distortion is mirrored in the primary electron shell so that the distance from the nucleus to the shell in the forward direction is smaller than the corresponding distance in the rearward direction. Higher energy electron shells are similarly distorted so that the distance between shell positions in the forward direction is less than the equivalent distance in the rearward direction. We now have an elongated atom with the nucleus displaced towards the upstream face.

If the atom in this distorted shape is excited so that it emits a "photon" the Zeron shell collapses over a smaller radial distance at the front than at the rear. However the collapse time is the same for the entire shell. The light wave is therefore projected backwards at a greater speed than it is projected forwards. In fact as shown in the diagram above, the light wave is ejected from the atom in the forward direction at a velocity $L = c-V1$ where $V1$ is the velocity of the atom through the Cosma. This is the ultimate heresy. A light wave radiating from an atom at less than the speed of light!

If it were not for the movement of the atom through the Cosma, the forward velocity of the light emission would be less than the rearward velocity and we would have different velocities of light in different directions. However the atom is traveling through the Cosma at velocity $V1$ in the forward direction. Add this forward velocity to the velocity of forward emission or deduct this forward velocity from the velocity of the rearward emission, and one gets exactly the same result in either direction! Due to this compensatory effect, light is emitted at the same speed forward or backward regardless of the velocity of the atom through the Cosma. Emissions at right angles to the direction of motion remain unaffected and in between emissions similarly result in a velocity of c of the light wave relative to the Cosma. We can therefore say that:

In respect of the velocity of emission of light from the atom, shape distortion and movement through the Cosma produce exactly compensating effects.

The light wave velocity through the Cosma turns out to be constant in all directions. Any movement of the atom relative to the Cosma or visa versa is effectively immaterial. We may further state that:

The speed of propagation of light from an atom is constant relative to the Cosma regardless of the movement of the emitting atom through the Cosma.

This is only half the story. What happens at the receiving end? It's well and good that the light wave travels at velocity c relative to the Cosma but what if the receiver is also moving through the Cosma. Surely then, if the receiver is traveling towards the light source, he will receive the light wave at a speed greater than c? Now for the distinction between facts and perceptions.

DIAGRAM 17

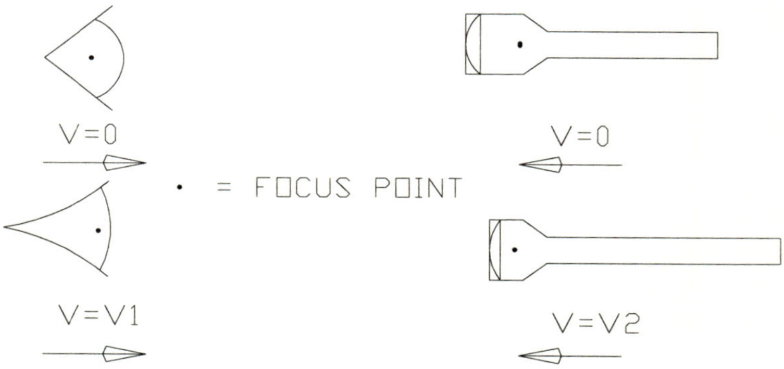

$V=0$

$V=0$

· = FOCUS POINT

$V=V1$

$V=V2$

Suppose a light source moves towards an observer at a velocity V2 relative to the Cosma and that the observer (receiver) moves through the Cosma towards the light source at a velocity v1. The matter of the light source is now distorted through its movement through the Cosma so that the emission of light from its atoms is at a velocity of c-v2. To this we add the velocity of the light source (v2) and we get the light traveling through the Cosma at a velocity $c-v2+v2=c$. The light wave travels at velocity c relative to the Cosma.

If the receiver were stationary relative to the Cosma, there would be no problem in deducing that the receiver would receive the light wave at velocity c. But because the receiver is moving toward the emitter at velocity V1, he receives the light wave at a velocity of $c+v1$ i.e. faster than the speed of light! Einstein at this point turns over in his grave. However the poor receiver is distorted and he will never know! No matter how he attempts to measure the speed of the incoming light, his distorted instruments or his eye will lie to him. He perceives light to be coming in at faithful old c! This is the perception. The truth is that light is pouring into his eye at a speed greater than c. Here at last is the truth about relativity. It is based on perceptions not facts.

Einstein's principle that the speed of light is constant relative to the observer is correct insofar as this is a perception. However as with all of Einstein's theorems, this is an "apparent" effect. In truth the speed of receipt of the light wave for an observer traveling towards a source at velocity v is $v+c$. He just can't measure the v component.

There is a very interesting outcome for all observers. As soon as the observer moves through the Cosma in any direction and at any velocity relative to the Cosma, shape distortion forever hides this movement from him. Thus the observer, whatever his movement, perceives the Cosma to be stationary!

Living in a stationary Cosma he would be quite entitled to assume that his "stationary" Cosma is the preferred reference frame and that he must therefore be in a very privileged position at the very centre of the universe! The problem is that he would undoubtedly have an argument with a friend traveling towards or away from him, who would also perceive a stationary Cosma! If we were subscribers to the "if we can't measure it, it doesn't exist" brigade we would now be able to enunciate the proposition that the Cosma is stationary relative to all observers. However in a more objectively real description we can state that the Cosma is apparently stationary relative to all observers.

This makes the Cosma the preferred reference frame for all observers who will "see" an apparently stationary and identical Cosma under all conditions. Here at last is the perfect reference frame. Everybody sees it exactly the same. The Cosma now becomes the preferred and universal reference frame! Stated concisely:

The Cosma is the Universal Reference Frame.

This is a truly astounding outcome, one that would have seemed impossible to achieve.

If you look at the outcome of Zeron theory relativity it is apparent that as far as perceptions are concerned, it is irrelevant as far as the observer is concerned whether there is a Cosma or not. This is precisely the result Einstein came to in his Theory of Relativity. However Einstein did not get to the reality and his theory remains one of perceptions only. Einstein equated perceptions with reality and then proceeded as if the Cosma (Aether) did not exist. And so it has been ever since. As we have now seen, Zeron theory, based on the existence of a Cosma, presents a radically different model and shows what is actually happening in both the micro and macro scale. That is progress indeed.

As we shall further see, the universal reference frame removes many of Relativity's paradoxes and illogicalities. We will be left with an objectively real and rational description of the universe. Such was one of the objectives of this book.

One plus one does equal two

How does the universal and preferred reference frame of the Cosma impact on the addition of velocities and special relativity? A simple thought experiment illustrates the difference between Special Relativity and Zeron Theory.

DIAGRAM 18

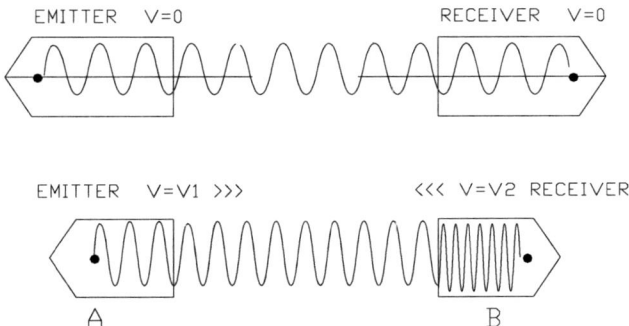

Assume two spaceships A and B are some distance apart and that each spaceship is stationary relative to the Cosma. Spaceship A is equipped with a light emitter and spaceship B with a light receiver. Both use the Zeesec as the time standard, the Zeesec being appropriately determined. Spaceship A is now accelerated towards B until it reaches a velocity of V1 relative to the Cosma.

We now arrange that A beams a monochromatic sodium sourced light towards B. The number of oscillations emitted from the sodium atom as the electron shell collapses is the same in all directions. However on the "compressed" (frontward) side of the shell, the linear ejection speed of the oscillations is lower than normal due to the decreased distance through which the electron shell collapses in the collapse time. This lower ejection velocity is the exact complement of the forward velocity of A through the Cosma so that the light emitted is emitted at velocity c. However because the number of oscillations in the emission is unaffected by atom distortion, and because the distance through which the electron shell collapses is less than normal at the front end, the frequency of the light wave is altered. We now have a light wave that emits at a normal speed, c, but with increased frequency. This "blue-shifted" light beam is directed towards B.

The God Particle

Blue shifted? If one looks at the spectrograph of a light source, one finds that the light beam is broken down into its colour components. Typically each of the elements has its own specific resonance frequency and this shows up on the spectrograph as a frequency peak. The frequency of sodium is 5.09×10^{14} Hz. If the light source moves towards us, this frequency increases and the light emitted by the sodium atoms moves towards the blue end of the spectrum. This is called the "Blue Shift" Conversely, light emitting objects receding from us get their light "red shifted". Red shift is a major tool in the astronomer's bag of tricks. Back to our thought experiment.

If B now observes the light from A, a "blue shift" (higher frequency) would be observed. The amount of the blue shift can be measured and B could correctly compute that A is approaching at V1. If B is now accelerated towards A to a velocity of V2 every atom in B is also shape distorted. The shape distortion exactly compensates for the forward movement so that any measuring instrument aboard B will measure the Cosma to be stationary. B Now observes the incoming light wave. What he sees is a light wave coming in at a velocity of c. However exactly the same sort of model applies to the receiver as the emitter so that the frequency is once more blue shifted. Both move towards each other at a velocity of V1+V2. If B were now to measure the closing velocity by means of this blue shift, they would correctly compute their closing velocity as V1+V2. If V1 and V2 were .8c each, the closing velocity would be 1.6c.

At this point Einstein once more turns over in his grave. He's having a restless time. The great illusionist has just had his greatest illusion uncovered. Using Einsteinian relative time the closing velocity would be measured as .98c. That's invalid. Why is it that he did not address the frequency issue? The only true measure of closing velocity of two objects is not light speed or apparent object speed, but blue or red shift. Had Einstein revealed this aspect of relativity, his theory would have been revealed for what it was, only a perception with little relevance to what is actually happening.

If now the captains of the spaceships decide they want to know how long it will be before they collide, they could arrange reflectors on A and B and measure the distance between them by timing the return of the light beam and dividing this figure by the closing velocity which they have correctly determined by blue shift. The answer would be valid and would them to take evasive action at the right time.

It might be thought that, by contrast, in Einsteinian relativity, B would measure the closing velocity to be a figure less than the speed of light and a nasty surprise would inevitably await. However Relativity Theory produces precisely the same answer by allowing time to be a variable. The distance as measured by a returning light beam, (measured in terms of altered Einsteinian time) will be less than it actually is and the division of the reduced distance by the reduced velocity yields exactly the same result! It can be seen from this example that while the relativity equations can be used as a calculating tool to produce mathematically correct results, they in no way reflect the actuality of the situation. The short description of the outcome of all the above is that

Zeron theory establishes the Cosma as the universal frame of reference and the Zeesec as the constant of time.

A train of thought

Let us now return and have a look at Einstein's most famous thought experiment. It's the one he used to develop the theory of relativity. A very long train travels along an embankment. The rear end of the train is A* and the front end is B*. A and B are two points on the embankment the same distance apart as the length of the train The train is traveling from left to right at a velocity V when it passes the two points A and B on the embankment

Diagram 19

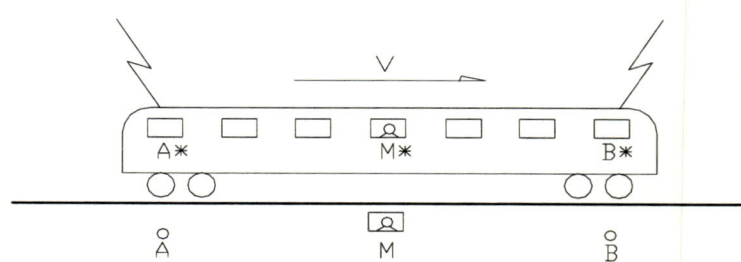

Now Mr. M is on the embankment exactly mid way between A and B. Mr. M* is riding on the train and has a seat exactly in the middle of it. As the front of the train B* passes B lightning strikes each end of the train simultaneously. M observes that the lightning has struck simultaneously at each end of the train because the light waves have the same distance to travel to him from either A or B. Not so for M*. Because he is in a moving train the light wave from B travels to him at a speed of V + c and the light wave from A travels to him at a speed of V − c. The logical result is that the light from B reaches him before the light from A so he sees the B flash first.

Now Einstein *knew* that both A* and B* always receive the light at velocity c regardless of their motion and would see the flashes simultaneously. In his view, the only way out of this dilemma was to make time a variable. Without going into the mathematics, if time for M* is not the same as for M then it could be arranged for both M and M* to see the flash simultaneously from each end of the train. Thus was Relativity Theory, the ugly child of Variable Time, born. Relativity was all to do with different frames of reference for systems moving relative to one another. Each frame of reference had its own dimensions and own time, which was considered to be a fourth dimension.

Einstein's equations *did* allow for contraction in length but it was applied on the macro scale to the entire frame of reference and was not applied at the atomic dimension. There was an inherent problem that prevented him from doing so. Had Einstein attempted to apply his equations at atomic scale, they would not have worked. The fundamentals of relativity were and still are deficient in this respect. His foreshortening equations only considered the "front window" view and were inadequate to cope with the "rear window" distortion that makes Zeron theory work. Let us see how Zeron theory solves the same problem of bodies moving relative to one another.

Assume, for now, that the embankment is stationary relative to the Cosma. By traveling at a velocity V each atom comprising the train is distorted. The distance from the nucleus to the electron shells is shorter in the direction of motion (front) than at the back. Lightning now strikes the train simultaneously at A* and at B*. The atoms at each end get very excited and they emit shells of light towards M*.

Now we have seen previously that the light leaving A* and heading towards M* does so at a velocity less than c because the distance from the nucleus to the electron shell in that direction is less than normal. The collapse therefore happens over a shorter distance than normal within the collapse-time of the shell and this results in a slower ejection velocity. The light leaving a B* atom and heading towards M* leaves at a velocity greater than c because the nucleus to shell distance is greater than normal and the collapse time is the same as for an A* atom. We also saw that this is an exactly compensatory effect. For A* light leaves it at a velocity of c – V. Add the velocity of A* and you get

$$c - V + V = c.$$

For B* light leaves it at a velocity of c + V. Deduct the velocity of B* and you get

$$c + V - V = c.$$

Just as for M, M* sees both flashes simultaneously because the light travels at the identical velocity c from A* to M* and from B* to M*. Simple isn't it? The key outcome is that Time has remained absolute and non-variable.

What if the embankment is not stationary relative to the Cosma? Well then the embankment shrinks in the direction of motion through the Cosma. The train shrinks even more and we get exactly the same compensatory effect. It's as if the Cosma is stationary to all observers regardless of their motion through it. It really is the Universal Frame of Reference.

There are always alternative ways to solve problems. Unfortunately Einstein chose the easy way out, a way that ever since has bedeviled physics.

Chapter 18

Revelations

Why light travels at velocity c through the Cosma. We derive the $E=mc^2$ equation from Newton's equation for kinetic energy. $E=m_z c^2$ is the quantum equivalent of Einstein's equation. We uncover the mechanism of increase of mass of the atom as it speed up in the Cosma.

Swift Zerons

The foreshortening/afterlengthening result of the modified Einsteinian equation gives us a method of deducing the velocity of a fast-Zeron in the Cosma. Now this is real progress. Suddenly we are about to deduce a concrete velocity for this totally elusive "virtual" particle. We will even go on to deduce the mass of the Zeron. It must be real!

In accordance with the above equations, at the speed of light relative to the Cosma, the dimension of the atom in the forward direction is zero. This means that the distance between the nucleus and the first or any subsequent shells is also zero. The rearward equivalent is a nucleus to shell distance, or a shell to shell distance, of $\sqrt{2}$ times the normal distance.

If we go back to the mechanism of light emission from an atom, we saw that the light wave is produced when the electron shell collapses to a lower energy level. The entire shell collapses in a specific time. Because the atom is distorted by movement in the Cosma, the distance through which the shell collapses on the front end is less than normal. The result is that the velocity of propagation of the light wave is reduced because there is fundamentally less distance of collapse within the same collapse time. This is the ultimate Einsteinian heresy. Here is a light wave being propagated at less than the speed of light! Hang on. All is not lost. It so happens that the slowing down of propagation exactly compensates for the velocity of the atom through the

Cosma. The end result is that the light wave is propagated at the correct velocity c relative to the Cosma.

The converse applies to the rearward direction. The velocity of the rearward propagation exceeds the normal speed of light by the same amount as the velocity of the atom through the Cosma. The two velocities added together result in a normal light velocity c relative to the Cosma. For v=c the distance between shells in the forward direction becomes zero and no light can be emitted (infinite frequency). The converse is true in the rearward direction. As the intershell distance increases the rearward emanating waveform can be pictured as "stretching" so that a lower frequency is emitted. At the limit of root 2 times the original static intershell distance the waveform becomes a straight line and no light can be emitted (zero frequency). These models constitute a highly rational overall model for light emission by a moving light source.

At the beginning of this book we stated that the Zeron had an intrinsic velocity of $\sqrt{2}$ c. Let us develop this further by falling back on Newtonian physics:

$$E = 1/2 \; m \; v^2$$

This is the equation for the kinetic energy of a moving body. That is the kinetic energy of any particle is 1/2 times its mass times the square of its velocity. Substitute the mass of a Zeron m_z for m and the velocity of the Zeron, $\sqrt{2}c$, for v. The resulting equation is instantly recognizable

$$E = m_z c^2$$

It's Einstein's equation for the equivalence of mass and energy! Now that we know that this equation can be derived from considerations of the Zeron, it is not surprising that there is an underlying rational Zeron model for the actual mechanism involved in this equation. The transformation of mass into energy is accomplished by the release of Zerons from the outer mantle of the nucleus of atoms into the Cosma. The nucleus loses mass and the increased Zeron population density in the surrounding Cosma results in the increased frequency of Zeron resonance within this region of the Cosma. This increased frequency is perceived as a higher energy level. Now, not only do we know Einstein's equation, but we know how it works!

Two principles important for the further development of Zeron theory arise from the above.

$$E = m_z c^2$$

is the underlying quantum equivalent of Einstein's equation. Here we have the first signs of the melding of relativity and quantum physics. The assumption that fast-Zerons have an intrinsic velocity of root 2 times c, produces a logical and explainable outcome. This assumption must be correct. Therefore the velocity of a Zeron in the Cosma (fast-Zeron) = $\sqrt{2} c$

The velocity of the Zeron underlies the speed of light.

Massive increases.

Einstein's equation for the increase of mass with velocity can be derived from the Lorenz transformation equations and the principle of conservation of linear momentum. The equation is

$$m = m_o / \sqrt{1-(v/c)^2} \quad \text{...................................... (1)}$$

You don't have to read or understand the equation. The important thing to understand is that mathematically, as the velocity tends towards the speed of light c, the mass tends towards infinity. The question is why? What actually happens? Let's look at the Zeron theory for answers

As derived previously, the shape distortion of any body moving through the Cosma at velocity v is,

$$L' = L \sqrt{1-(v/c)^2} \cos \Phi \quad \text{..................................(2)}$$

The only way that a body, an atom or a particle can increase mass is by increasing its Zeron content. A nucleon that is stationary relative to the Cosma absorbs and emits Zerons at the same rate. Except for minor short-term fluctuations, the nucleus' Zeron count stays constant. But as soon as the nucleus or atom is

The God Particle

accelerated in the perfect fluid of the Cosma, this balance is disturbed. The atom/nucleus becomes shape distorted and at low velocities, marginally more Zerons are accumulated from the front than released in any other direction. This accumulation process only happens during the acceleration process. As soon as the acceleration stops the atom/nucleon becomes stable once again and moves through the Cosma without any resistance. However the shape distortion that was extant at the end of the acceleration becomes the new stable configuration and it is found that the Zeron count has increased. The mass increase in equation (1) can be explained by analysing equation (2).

If equation (2) is plotted for different values of Φ and v, we get a model of the shape distortion of the electron shell, and by implication, of the nucleus. The effect at low velocities is exceedingly small because the shape distortion of both the nucleus and/or the electron shell is also very small. The distortion is in the form of a slight flattening on the front face and a larger but still insignificantly small elongation in the rearward direction. As equation (1) indicates, the effect is not linear with speed and there is a pronounced increase in the mass gain as the speed becomes closer to the speed of light. The reason for this effect becomes clear if equation 2 is plotted for v= .8c, .9c, .95c and c. as shown below.

DIAGRAM 20

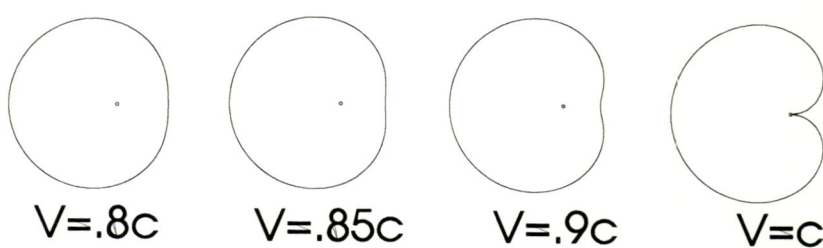

V=.8c   V=.85c   V=.9c   V=c

The above plots represent shape distortion of the Zeron electron shell. However this resonance shell is the exact mirror of what is going on in the nucleus, as it is primarily the distortion of the nucleus that sets up the distorted shell. As the velocity gets to higher values the shape distortion becomes more significant. At

.8c the overall diameter can be perceived to have increased slightly behind the 90-degree line and there is a noticeable flattening of the leading face. At .85c the flattening has become more pronounced and the frontal area has increased further. At .9c a "dimple" has formed on the front face forming a trap for fast-Zerons during acceleration. The number of excess Zeron captured becomes more significant. This dimple becomes more and more pronounced as the speed increases until at the speed of light the shape has become a flattened sphere with a curved re-entrant cone at the front. The cone terminates sharply at the centre of mass of the nucleus.

Here is the perfect mechanism for the absorption of Zerons. As the particle is accelerated so ever more Zerons are trapped than can be dissipated and the nucleus increases in mass. Look at the last plot. It's a perfect funnel. It channels Zerons right into the heart of the nucleus.

Here at last is a good visualisable model of how the accelerating atom gathers mass. It also explains why this mass increase is not linearly tied to the velocity, but increases exponentially as lightspeed is approached.

There is an interesting side-play to this model. For as long as the atom is held at lightspeed, Zerons will continue to be funneled into the nucleus with no chance of escape. The nucleus will get ever more massive and theoretically would go on its way trying to achieve infinite mass by absorbing all the Zerons in the universe! Of course this would take infinite energy which is not available. However if all the available energy in the universe were to be applied to this task, the nucleus would indeed swallow everything. This is another way of saying that the conversion of the all the energy of the universe into mass could result in the formation of a single super-nucleus of totally condensed matter. This is precisely the inverse of the Zeron model for the Big Bang!

To summarize, the increase of the mass of a body is due to acceleration (not velocity) through the Cosma that results in an accumulation of Zerons in the nucleus. The maintenance of this increased mass is due to the velocity through the Cosma, which stabilizes the shape distortion ensuring the retention of accumulated Zerons.

Chapter 19

Describing a Planck

By examining the relationship between the energy of a photon and the kinetic energy of a Zeron, we find that Planck's constant is the energy of a fast Zeron. We then derive the mass of the Zeron and find it to be many orders of magnitude less than the electron. We live in an extremely dense "soup" of these Zerons that we are totally unable to detect.

Once more we beg your indulgence while we examine yet another equation. It's an important one, which even today is poorly understood.

Planck's puzzling constant h, derived from the graph of black body radiation, and one of the foundational constants of Quantum Physics, has always been thought of as some sort of manifestation of the underlying "grain" of matter. The constant is exceedingly small, and has the strange dimension of erg-sec. It comes into play in virtually all quantum calculations. What this constant (or its companion h/2pi) represents, if anything, has always been a mystery. Zeron theory provides a rational explanation.

We have seen that light emanates from an atom as the Zeron electron shell collapses from a stable allowable radius to the next smaller allowable radius shell. In the process it sends out an expanding shell of Zeron momentum energy that is the matrix from which the photon is formed. The photon consists of a series of Zeron impact waves and contains some 10p5 to 10p6 oscillations.

One equation that incorporates Planck's constant is for the energy of a photon.

$$E = h \times f \quad \text{...................................... (1)}$$

Where E is the energy of a photon, h = planks constant and f is the frequency of the photon. The Zeron interpretation of the photon is that while, in a general sense, it is not the Zerons that are translated by the light wave but rather it is the Zeron momentum that is translated through the Cosma, there is, for each light wave oscillation, the movement of Zerons by one "pitch". The foremost

Zeron in the path of the photon wave dumps its energy on the absorbing surface by impacting on the surface. It loses its kinetic energy through absorption into the electron shell of the atoms on the surface. The total energy of the "dumped" Zerons is the sum of the kinetic energies of the Zerons. Now the total number of Zerons dumped is the frequency of the light packet times the period for which it is transmitted, i.e. f x secs. Going back to Newton, the total kinetic energy of the Zerons can be expressed as

$$E = 1/2\, m_z v_z^2 \text{x f x secs} \dots\dots\dots\dots\dots\dots\dots(2)$$

But the velocity of the Zeron is $\sqrt{2}\, c$ where c = the velocity of light. The energy dumped by the light train is therefore

$$E = 1/2 \times m_z \times \sqrt{2}\, c \times \sqrt{2}\, c \times f \times \text{secs} = m_z c^2 \times f \times \text{secs} \dots (3)$$

We have therefore two equations (1) and (3) for the energy of a photon. Equating them gives

$$h \times f = m_z c^2 \times f \times \text{secs}$$

Dividing both sides by f

$$h = m_z c^2 \text{ secs}$$

But the dimensions of h are erg.secs and if we measure the photon energy in ergs, we get:

$$h\,(\text{erg secs}) = m_z c^2\,(\text{erg secs})$$

Dividing both sides by erg x secs

$$h = m_z c^2 \dots\dots\dots\dots\dots\dots\dots (4)$$

Does this look familiar? Compare this with

$$E = m c^2$$

Describing a Planck

Planck's Constant h turns out to be the energy of a fast-Zeron.

Planck's Constant is indeed a measure of the underlying grain of matter. It was just that nobody knew what sort of grain. Equation (4) is the quantum equivalent of the macro expression for Einstein's equation for the equivalence of mass and energy. Here once more we see the melding of macro Physics (Relativity etc) and micro physics (Quantum mechanics). It's a most satisfactory outcome.

From small things......

It seemed highly unlikely that one could derive the mass of an undetectable particle, but, as promised, that is exactly what we are about to do.

While it was always clear that in order for the Zeron to produce "Zeron pressure" on a nucleus, it would have to be many orders of magnitude smaller than the nucleus, the incredible minuteness of the Zeron still came as a bit of a surprise. I now know how the early physicists felt when they discovered that the atom only had a very small speck of matter at its centre and was seemingly mostly empty space. For the sake of the scientific community and those interested in convincing themselves of the rather surprising result, the derivation is given below. If you want to take my word for it, skip to the answer.

From equation (4) above, we have

$$m_z = h/c^2$$

Substituting experimental values for c and h we get

$$m_z = 6.63 \times 10^{-27} / 2.995 \times 2.995 \times 10^{10} \times 10^{10}$$

Zeron mass = 7.39×10^{-46} gms!

This is an *exceedingly* small mass. Even the tiny electron is composed of a great number of Zerons namely

The God Particle

$$9.11 \times 10^{-28} / 7.39 \times 10^{-46}$$

1 Electron mass = 1.33×10^{18} Zeron masses!

To drive the point home there are 1,330,000,000,000,000,000 Zerons in an electron! No wonder the Zeron is unlikely to be detected.

The unimaginable minuteness of the Zeron lends great credence to the mechanics of the theories so far put forward, especially that of the ability of the Cosma to exert pressure on the atomic nucleus. At the same time the Zeron represents a previously unknown and unimagined class of matter whose size is difficult to comprehend. In rough terms, the Zeron is to the electron as the electron is to a tennis ball. The dimensions of the Zeron, its velocity, and its physical properties make it clear why the Zeron will probably never be subject to direct detection by instruments which in themselves are composed of the enormously larger atoms. Incredible as it seems we appear to exist in this immensely dense Zeron "soup" which we are totally unable to detect! This soup is a different sort of Higgs Field, a postulated all pervading field consisting of Higgs Bosons. The conventional scientists were right in the postulate but wrong in their assumptions about the sort of field. Our field is the Cosma and our God Particle is the Zeron,

Chapter 20

The Truth Will Out

We examine the nuts and bolts of the flow of electricity through a conductor including the mechanism of superconductors. How Gamma rays are produced by the nucleus of an atom and confirmation of the strange properties of a muon atom. How resonance in the nucleus is necessary for stability. We compare Beta particles and electrons and reinforce the concept that electrons rarely exist in nature. We redefine the concept of "charge". Positive and negative charges have different rates of resonance. How positrons are produced. "Charge" is a manifestation of forces due to resonant effects and Zeron pressure and has no intrinsic existence as a discrete force or property of nature.

Electrifying flow

Having dealt with the nuts and bolts of Zeron theory, we now come to the part of the book that looks at various physical phenomena and seeks to explain them in terms of Zeron fundamentals.

The first of these is electricity. What is electricity? If, as we saw previously, the electron only exists when we *make* it exist, what actually flows through a wire when we pass current through it? Why *does* a current heat a wire? Why *does* a higher temperature wire have higher resistance? All these effects are described in conventional physics in terms of empirical formulae. Zeron theory describes the underlying mechanism.

The preferred state of the electron is when it is dispersed within the concentration of Zerons of the Zeron-electron shell. In the normal state the electron does not exist as a discrete particle. Within a conductor are many atoms with their electron shells each, so to say, minding its own business. The introduction of Zerons/electrons to one end of the conductor results in a net Zeron pressure within the conductor. Providing that there is somewhere for the electrons/Zerons to go at the other end, an electric current will be observed to flow as the electrons/Zerons move along the conductor. However it is a mistake to visualize a current flow as the flow of discrete electron particles from one end to the other.

There are two simultaneous effects when electricity is pumped into the end of a conductor. The first is that any Zeron pumped into one end of the wire ejects a Zeron at the other end. These two events do not take place simultaneously. The impulse that makes it happen travels along the wire at the speed of light. The second effect is similar to the process of the emission of light by an atom. The short version of what actually happens is that as Zerons are fed into the end of the conductor, they meld into the Zeron shells of the atom creating both a surplus and a directional pressure. Electrons are formed at the nodes of the shells and these "bleed" off each atom and are propelled along the direction of current flow until they are re-amalgamated into the shells of subsequent atoms. This formation of electrons is entirely man-made. Driving current through a wire is not a natural phenomenon. The premise that electrons rarely exist in nature is upheld.

The detailed description of the mechanism involved in the flow of electrons through a conductor is as follows. It is much the same process as for the emission of light. The shell of an atom within the conductor finds itself in a denser Cosma due to the ingress of Zerons and is suddenly out of synch with the Cosma. It expands to get in synch, only to re-collapse as the nucleus also gets into step. The shell collapses, but atoms within the conductor are constrained and instead of emitting a light ray, an electron droplet bleeds off, removing surplus Zerons and leaving behind a stable atom. The electron is impelled in the direction of current flow and gets re-absorbed by an adjacent shell. This shell, in turn, goes through the same procedure. This results in a chain reaction that is passed on down the conductor. This movement of electrons is quite slow. On the surface of the conductor, the atoms are unconstrained. These atoms are stimulated in the same way as the internal atoms but being unconstrained they emit radiation instead of electrons. We perceive this output of energy as heat or light. The flow of electricity can therefore be conceptualized as the successive formation and absorption of electrons from the Zeron electron shells of the material of the conductor.

If the flow of Zerons lies at the root of the flow of electricity, why do we need the above electron model? Cannot the Zerons simply flow from one end to the other? If the electricity flow were simply the flow of disassociated Zerons, this would happen in a loss free manner and it would not be possible to construct a model that would result in the production of heat and light. Without the sequential formation and absorption of electrons we would observe that electricity flows through a conductor without loss. This would represent zero resistance, not a normal observation.

Conductors generally have higher resistance at high temperatures. When you switch on a light bulb, there is, for an instant, a high current surge. This is usually when the light bulb blows. As soon as the filament has attained operating temperature, the resistance increases, and the current drops to normal values. At the Zeron level, the higher the temperature, the higher is the energy of impinging Zerons. Higher energy Zerons form higher energy shells that are more stable. The process of formation and re-absorption of electrons is impeded. We observe a higher resistance to the flow of electricity. Once more Zeron theory provides a visualisable model.

What about superconductors? At a few degrees above absolute zero, certain materials exhibit zero resistance. Not very low resistance, but zero resistance! A current induced in a super-conductor closed circuit will flow forever. Conventional physics has correctly identified the phenomenon as having to do with resonance. Zeron theory provides a somewhat more detailed explanation. What has happened is that by reducing the conductor to a very low temperature the atoms and their Zeron/electron shells are no longer disturbed by the anomalies such as occur at higher ambient temperature. In the absence of bombardment at sufficiently high energies to disturb the system, the whole character of the conductor changes. The atoms and their shells "get into step" and start to assume a single resonant frequency. The shells of all the atoms in the superconductor are now like the resonant shells of a single atom.

Atomic shells if left to their own devices, will resonate forever because the ingress and egress of Zerons into and out from the atom and its shells is loss-free. This loss free system is created from resonance within the shells and the nucleus. The superconductor behaves like a single giant atom. Zerons can enter and leave and electrons can form and be absorbed loss-free. As a result, the superconductor has zero resistance and a current induced in a closed coil or loop will flow forever.

Radiant Gamma

One of the more compelling aspects of conventional nuclear theory that lends credence to Zeron theory is to be found in the characteristics of compound nuclei. This is especially true of the model for the emission of "rays" from the nucleus. To recap, the nucleus is represented simply as an agglomeration of Zerons held together by Zeron pressure. The proton, neutron and other elementary particles are all agglomerations of Zerons that under favorable circumstances coalesce to form the nuclei of the many types of atoms.

Gamma rays are known to be similar in character to light rays. They are so-called "electromagnetic" waves of a transverse nature just as light waves are but have much higher frequencies, typically 10^{21}Hz whereas light waves have a frequency of about 10^{15}Hz. Gamma rays have therefore a frequency of about a million times the frequency of light waves. If we examine the dimensions of the electron shell relative to the nucleus of say a hydrogen atom we find about the same ratio. Gamma radiation is generated in the same manner as light but emanates from the nucleus instead of the electron shells. In fact gamma rays are a strong evidence for the existence of resonance shells within the nucleus. The nucleus is composed of low velocity Zerons. All the characteristics of the high velocity Zerons with their electron shells are replicated within the nucleus. In fact the fuzzy outer mantle of the quantum model of the atom's nucleus is the outer resonance shells of the slow-Zeron nucleus. The emission of gamma rays is based on exactly the same principles and mechanism as for the emission of light.

Diagram 21

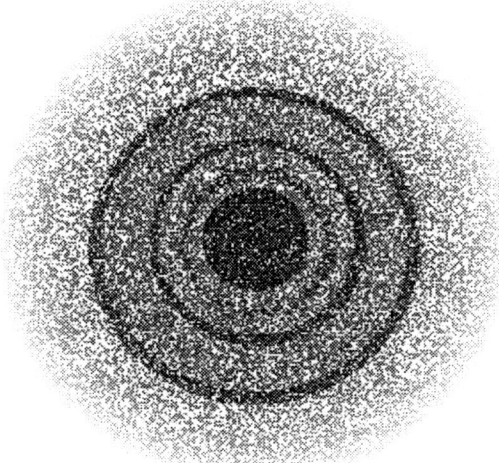

Nucleus showing shells and "hard" nucloid

The process for the emission of Gamma rays is as follows. It is known that a nucleus can be raised from its "ground state" to a higher resonant state. This happens in a quantum manner just as for the electron shell. It is the outermost shell of the particular nucleus that is involved. An input of energy (Zerons count in the surrounding Cosma increases) causes the outermost shell to jump to the next larger "allowable" radius. A return of the nucleus to its ground state follows as the nucleus "catches up" with the Cosma and the outer shell returns to its original radius in a small but finite time. This results in an outwardly traveling spherical shell of Zeron momentum. This is interpreted by the observer as a gamma ray.

The presence of resonance shells within the nucleus implies that there is a nuclear "de Broglie" type wavelength operating within the nuclear shells. The process of excitation and de-excitation of the nucleus would not change its atomic number or any of the other characteristics of the nucleus. These characteristics are in agreement with conventional scientific observation

An interesting special case of the similarity of electron shells and nuclear shells is to be found in the Muon atom. A Muon atom is an atom in which an electron has been replaced by a Muon, another type of particle. The electron shell of such an atom is 207 times smaller than a conventional atom. For elements with a high number of particles in the nucleus, the Muon "orbit" lies inside the nucleus! There is no explanation of this strange outcome from conventional physicists. It is difficult to conceive how this could be possible.

Zeron theory to the rescue! In the Zeron model such an occurrence presents no difficulty at all. The electron shell could be superimposed on the nuclear shell or be interposed between them without interference. This is in accordance with the well-known principle of superposition of wave patterns in which many wave patterns can be superimposed on each other without interaction between them. The upshot of the previous section is that it may be said that:

Nuclear shells are the slow-Zeron equivalent of Zeron electron shells.

"Resonance" particles don't resonate

Some hundreds of so-called "elementary" particles have been identified as a result of atom-smashing experiments. Conventional wisdom has it that there is a resonant effect associated with these particles. They have accordingly been dubbed "resonance particles". These are generally very short-lived decaying in 10^{-21} to 10^{-18} secs. This area of research appears to point to a general resonance phenomena in basic matter. This is well supported by Zeron theory that proposes a model of matter that is entirely dependent for its existence on a resonant state of the Zerons from which it is composed. However the reason that so-called "resonance" particles are not stable is because these particles do not in fact resonate - they simply have an internal vibration rate that is out of step with the Cosma. Only when the particle resonates is it protected from decay from the bombardment of fast-Zerons from the Cosma. Not being resonance protected, these particles decay very rapidly.

Even the Neutron in isolation is not a stable particle because it cannot resonate on its own. However once melded with a proton, or split into a proton and an electron, the resulting nuclei resonate and are once more stable. To achieve any sort of stability, a particle must be resonating in step with the Cosma. This can only happen if a standing wave configuration is achieved in the particle.

Beta decay

Certain types of nuclear decay produce Beta particles. Beta particles are identical to electrons. The mode of formation of a Beta particle is identical to that of the electron. It "bleeds" off from the electron shell like the drops of water vibrated from the surface of water subjected to vigorous vibration. In radioactive decay, the bleeding is a bit more vigorous. The electron is projected away from the atom at some speed and is not bound into orbit around the atom. We know from observations of Beta particles in cloud chambers that Beta particles are unstable and disappear within a short distance of the emitting atom. If the beta particle is identical to the electron, then we may deduce that electron particles are also unstable. Therefore the proposition that electrons indefinitely orbit nuclei is untenable. This adds evidence to the proposition that electrons rarely if ever exist in the natural world.

It is held that Beta particles also show up with a positive charge. In fact this positive Beta particle is nothing like a normal Beta particle. Beta is really a misnomer. It should be called a Positron. Positrons have the same mass as a beta particle but the make-up is quite different as will be shown in the following section where the difference between positive and negatively charged particles is discussed.

Charge!

Few concepts are so entrenched in physics as the concept of "charge". The concept of two different kinds of charge, "positive" or "negative" comes from at least 200 years back when it was

The God Particle

discovered that glass rubbed with silk and hard rubber or ebony rubbed with fur were found to have two kinds of charge which either attracted one another, (+ and -) or repelled each other (+ and + or - and -). Charge is now considered to be an intrinsic property of matter and one cannot get very far in physics without reference to this property. Strange then that there appears to have been no effort to uncover what "charge" actually is. Stranger still that modern physics is still based on glass, silk, fur and ebony!

Zeron theory does not support positive or negative charges as separate entities but rather considers these as two variations of a single phenomenon.

The concept of positive charge can probably best be explained in terms of the difference between a Proton and a Neutron. It is known that the Neutron is slightly heavier than the Proton. The physical difference in terms of Zeron theory is that the Neutron like the Proton is an agglomeration of Zerons but simply contains additional Zerons. However these extra Zerons dramatically change the character of the nucleon.

The Neutron is a non-resonant body. Like a cracked bell, the Neutron refuses to "ring'. Any Cosma-induced resonance is quickly damped by the self-canceling effect of out of synch waveforms in the nucleon shell structure. No standing wave resonance is possible. The Neutron does not resonate, does not affect the surrounding Zerons by producing its own resonant field and is therefore "neutral". In its free state its lack of resonance makes it vulnerable to the effects of the surrounding Cosma and in isolation, it quickly decays into a Proton and an "Electron". The Neutron is considered to have no charge. In general therefore

A particle without charge is a non-resonating agglomeration of Zerons.

By contrast, the physical shape and Zeron content of the Proton permits it to resonate when stimulated by the impact forces from the surrounding Cosma. It "rings" like a bell. Its resonance affects the surrounding Cosma and it is this characteristic that we perceive as "charge".

A charged particle consists of a resonating agglomeration of Zerons.

What about positive and negative charges? Once again we must look to the proton and "electron". If the proton resonance represents the "positive charge" then the electron resonance represents the "negative charge". How does the negative model differ? Very simply the proton is a relatively large chunk of matter composed of slow-Zerons. This represents the positive model. The electron is borne out of resonating Zeron shells. Electron shells are composed of fast-Zerons as is the electron and this represents the negative model. The essential distinguishing difference between positive and negative is the rate of resonance.

The rate of resonance within a particle determines whether it represent a positive or negative charge.

Why do like charges repel each other and dissimilar charges attract? What could possibly make two protons repel each other? We have seen that it is resonance in a nucleus that protects it from attack by the Cosma and makes it stable. In exactly the same way that resonance repels the attack of the high velocity Zerons in the Cosma with which it is in tune, resonance resists the approach of another resonating body that has the same resonance. We saw previously that an atom has many shells theoretically extending to infinity. These shells set up a quantum field. It is the interaction between two such identical resonating fields that we perceive as repulsion of one charged particle by another similarly charged particle. As is to be expected the repulsion reduces as the square of the distance between the particles. This is in accordance with the strength of the quantum field that also reduces in this manner. So much for the positive/positive or the negative/negative model to which exactly the same conditions apply.

If two resonances are not matched there is no repulsion effect. Why then is there an attractive force between positive and negative? The short answer is that there is *no such attractive force*. In the absence of a repulsive force, the normal out-to-in force due to fast-Zeron pressure and interference effect comes into play. It's the longer distance version of the strong nuclear force.

What about the rather rare particle called the positron? This has a mass equal to that of the electron but has a "positive" charge. There were historically, good theoretical grounds for believing

The God Particle

that such a particle existed although none had ever been observed.

Then an electron was seen to trace a curved path in a magnetically influenced bubble chamber. Nothing strange about that until it was realised that the electron's path curved the wrong way. The positron had been discovered. The technique for making positrons is now well known. It is instructive to examine how a positron is "manufactured". Physicists use gamma ray particles to bombard atoms. An electron and a positron are produced as a pair. The key to understanding the positron is that it cannot be produced in a vacuum. It needs an atomic cloud "to absorb the rebound" (or so the physicists say).

In actuality what is happening is that the Gamma ray is pumping energy into the atoms causing them to over-resonate. (This cannot happen in a vacuum with no atoms) In response, the nucleus of one of these atoms throws off the Positron and its Electron shell throws off an Electron. The difference between these two particles is that the "Electron" is composed of fast-Zerons and the Positron, which comes from the nucleus, is composed of slow-Zerons. They have the same mass. The Positron as for the Electron has a very short lifespan because they both vibrate at a frequency at which a standing wave formation cannot be formed. Neither achieves resonance, and they are soon dispersed into the Cosma. This dispersion is perceived as the release of energy. The Positron and the Electron are really freak particles. In conclusion, therefore

"Charge" is a manifestation of forces due to resonant effects and Zeron pressure and has no intrinsic existence as a discrete force or property of nature.

Chapter 21

Circular Arguments

"Spin" is an ill-defined property of matter in conventional physics. It has no real meaning. In Zeron theory, spin is a manifestation of the dynamic shapes taken by a nucleus as it goes from spherical to oblate shape to an elongated necked shape and back to spherical. Explanation of why a nucleus with a spin of one half has to spin twice to look the same. We create models for magnetism and explain why magnetism can attract or repel. The splitting of the atom is re-examined and we identify the source of the energy resulting from this event.

Spin Doctors?

The theories concerning spin in conventional physics are both complex and very fully developed. The different categories of spin can be successfully used as tools to categorize different particles of matter and their physical properties. Spin 1/2, 1 or 1 1/2 refers to a number of complete rotations of a nucleus in a given time but the definition is hazy and the model non-existent. What exactly spin refers to is not known.

Quantum Theory gets it partially right. "Spin" in quantum physics is not regarded as a physical rotation or orbital motion, but as "an intrinsic property of matter". As such "spin" might have been given any other nominal name such as "charm" or "colour", names for properties of the quark. "Spin" is therefore somewhat of a misnomer but as we shall see, the term chosen for this property is not so far from the truth.

A number of basic models of nuclear resonance can be built up from Zeron theory. One model has a spherical nucleus simply resonating internally but not shape resonating. A second model has the nucleus resonating as before but changing in shape from sphere to oblate spheroid, (rugby ball) and back again through sphere to oblate spheroid and so on. A third model is similar to the above but the extreme distortion is in the shape of a prolate spheroid (flying saucer shape).

Neutron scattering experiments seem to indicate that most nuclei are not spherical but are one of the varieties of spheroids. It is surmised that the neutron scattering experiment is actually picking up one of the dynamic spheroidal forms as the nucleus goes through its shape fluctuations.

Like a free drop of water floating in a space capsule, there would at first glance seem to be no valid reason for a nucleus to be anything else but spherical. After all, Zeron pressure comes equally from all sides. Two factors mitigate against the perfectly spherical nucleus. While on average, the Cosma is indeed homogeneous, at very short time-scales, there are many aberrations that cause differential pressures. Secondly we are dealing with a totally undamped system and any induced shape distortion effect will carry on forever.

In a theoretically isolated volume of space, the Zeron impact on a nucleus would be equal from all directions and the nucleus would be perfectly spherical. However there is certainly no place in the universe where such conditions exist. "Empty" space is criss-crossed by all manner of energy transfer that upset the homogeneous nature of the Cosma. Light rays and all types of radiation, from gamma to long infrared, pass through the Cosma at every point in the universe. Ultra high-energy cosmic rays and other high-energy phenomena regularly transverse the Cosma in a loss-less manner. All these patterns are imposed on the amorphous Cosma. Now in most cases, these energy trains pass through losslessly and they leave no permanent imprint on any atoms that are in the vicinity. But occasionally, and for very short periods of time, there is a combination of these waves at a particular place in space to form a resonance node. Resonance nodes, as we have seen in the formation of the electron, can interact with matter. If there is a nucleus at this point it is going to get a jolt and shape distortion will be initiated. Once shape distortion of the nucleus has been set up, it will shape-resonate forever. Over the very long time that the universe has been in existence, every atom in the universe has been exposed to this process. Shape resonance is universal in the nuclear world. It might be asked why different nuclei of the same element do not have different shape resonance.

Surely all nuclei have not been exposed to identical jolts. No they haven't, but shape resonance for any particular nucleus has a limiting value. Shape fluctuations will build up until the out-to-in force imposed by the Cosma is greater than the dynamic forces within the shape fluctuating nucleus. It's a sort of tug-o'-war with one side overcoming and then the other. Nobody ever wins. At the limit the nucleus will fluctuate in shape at a frequency determined by the number of Zerons in the nucleus and will continue to do so indefinitely.

This has been a rather long introduction to the concept of spin. It has been necessary because as we shall see, spin is a direct outcome of this shape resonance.

DIAGRAM 22

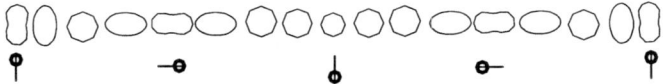

OBLATE ATOM SPIN

Consider an atom in a state of oblate shape resonance. At maximum extension the major axis points in a particular direction. It is possible that the formation of the major axis of the oblate spheroid could form in a progressively offset manner. Look at a globule of oil breaking free in the popular "Lava Lamp" Even in this highly damped environment, the globule shape resonates a few times and the major axis does not occur at the same place each time. Say for instance that the first major axis of an oblate nucleus forms in a particular direction which we specify as 0 degrees. If the following major axis appears it might be at an angle of 30 degrees and so on. If we took a snapshot at each re-appearance of the major axis and turned it into a movie, then we would have a moving picture of an oblate spheroid rotating about a vertical axis. It would appear to be spinning. This model neatly complies with both the conventional spin model and the quantum spin model.

The nucleus does not spin, but appears to spin. This is why scientists have realised that spin isn't spin but is "an intrinsic property of matter". How close we get with conventional physics, but never get to crack the reality barrier!

The illustration above demonstrates another counter-intuitive outcome from conventional physics. At a spin of 1/2 it is said that a nucleus has to rotate through 360 degrees *twice* to appear the same! Surprisingly this apparently nonsensical statement has some truth in it.

In the diagram above we have shown a nucleus going through shape distortion from the extreme distortion in which it necks to oval to spherical and back again. The sequence is repeated twice but at the end of the first quarter cycle the atom elongates vertically to the page. The lollipop below shows the actual "rotation". The whole sequence is for one 360 degree rotation. However if one looks at the sequence of distortion is could appear that the nucleus rotates 360 degrees twice! It might then be said that two rotations of the nucleus are necessary for it to return to its original state. As with much of conventional physics, appearances are everything!

Spin is an extremely complex phenomenon incorporating left and right spin, half and whole number spins, and spins having an angular momentum of ½ up to 20. Half spins have a different characteristic to whole spins. Each value of spin has its own characteristics. The complexity goes on forever. A book can be written on the subject. We will have to leave it to those interested in re-developing spin theory with the concept presented here. The foundation has been laid.

When is a magnet?

It might be wondered why the explanation of one of the prime forces of nature, magnetism, has been left so late into the book. It is because magnetism is intimately bound up with the concept of the "spin" of nuclei. In common with the other basic forces of nature magnetism is a manifestation of a differential Zeron pressure.

Circular Arguments

If one examines any illustration of a magnetic "field" of a bar magnet, one finds a pattern of "force lines". For the bar magnet these appear to emanate from the North Pole, curve through space, and re-enter the South Pole. In the early days of physics, these lines were traced out by means of a compass needle, but what they represented was a bit of a mystery. Faraday was the main experimenter in the mid-1800's and his friend Maxwell being a consummate mathematician instantly recognized the field patterns. He derived his famous equations on this basis. These equations still stand unchallenged today nearly 150 years on, but they remain just equations. There is still no satisfactory physical model of magnetism. Many models of magnetism have been suggested. They all seem to recognise the inherent rotational component of the phenomenon. Zeron theory joins the majority view. The only model that matches the physical characteristic of a magnetic "field" is one of a vortex of Zerons. If we look at the North Pole head on the vortex would be seen to be spinning in a clockwise direction. The vortex spins in the same direction along the whole bar magnet.

DIAGRAM 23

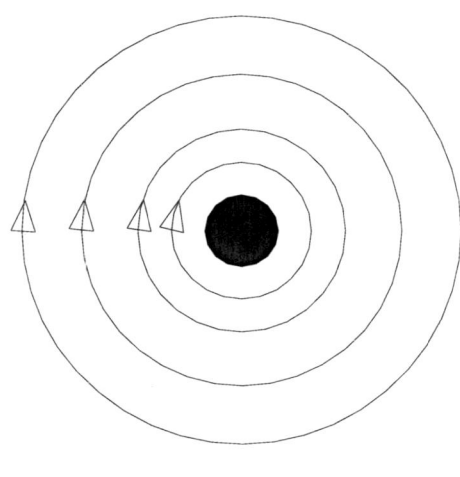

North Pole

Bringing the N and S poles of two bar magnets together results in the vortexes produced by both bar magnets spinning in the same direction. Because the two vortices reinforce each other, there is now a greater degree of order within the Zeron population between the two magnets than there is on the remote ends of the magnets. In other words, many of the Zerons that would normally be impacting the ends of the magnet are now coerced into the vortex and do not impact on the ends of the magnet. There is now an imbalance of forces on the bar magnets with greater Zeron impact on the outer faces of the bar magnets than on the inner faces. This imbalance forces the magnets towards each other. We perceive this as "attraction". In fact it is pressure from the opposite sides of the bars.

Likewise if two N poles are brought together, the opposing vortices create greater randomness between the magnets than at the ends. The Sum of Zeron impacts between the inner ends of the bar magnet is now greater than the outer ends and we perceive a "repulsion". It's that simple. Here at last is a physical explanation of magnetism.

But hang on! Why *do* the Zerons form a vortex? It's to do with Spin. In the previous section we saw that the major axis of a shape-resonating nucleus can occur with a rotating pattern. The conventional physicist recognizes this property as spin. The so-called "ferromagnetic" materials have the property of having the major axes of many of its atoms aligned in the same direction. Once set up, these atoms have the ability to "latch" in that orientated position. The effect of this is to set up a rotation in the surrounding fast-Zerons of the Cosma. This all makes sense if it is remembered that the "permanent" magnetism of a magnet can easily be destroyed by simply heating the magnet. By doing so, enough random Zerons are introduced into the material of the magnet to upset this ordered alignment and the magnetism disappears. It also makes sense that when re-magnetizing a permanent magnet it helps to lightly tap the magnet during the process. The physical shock encourages atoms to get aligned by jolting them.

What about electro-magnetism? Surely there can be no simpler model. A coil of wire with Zerons running along it will set up a vortex in the surrounding Cosma. Where there is a magnetic core material, the circulating Zerons in the winding will strongly orientate the atoms in the core to make it the equivalent of a bar magnet. Perhaps there's another Maxwell out there who will derive Zeron-based equations for this model?

Getting more compound

We now go back to shape resonance for a while to pick up some interesting explanations for well-known phenomena. There is one situation where the Shape Shifter tug-o'-war team triumphs over the Cosma team. It occurs spontaneously in over-heavy atoms created in the laboratory.

The nucleus of the Hydrogen atom consists of a single proton. Nuclei range in size from this, the lightest element, to the very heavy elements such as Uranium, which has 235 or 238 nucleons in its nucleus. It is possible to manufacture even larger stable nuclei. However, over-large nuclei have short life spans. If the mass of the agglomeration constituting the atom is too large, the confining pressure of the Cosma is insufficient to prevent over-distortion. Too many Zerons in the nucleus, and a neck forms as the rugby ball shape over-extends because of the extra mass of either half. If the neck becomes too accentuated the pressure from the Cosma narrows the neck further until the nucleus is divided. Without exception such over-heavy nuclei have extremely short life spans. There is no explanation from conventional physics as to why this is so. Zeron theory provides the model. It is not only the over-heavy nuclei that undergo this change. Some nuclei such as uranium sit on the very edge of instability and can be induced to split, releasing large amounts of energy.

Is this all fanciful? I think not. The following diagram was taken from a series of photos of an experiment in which a water drop suspended in oil is pulled apart by induced electrostatic charges. The photos constituted the model for the splitting of the atom for the production of the first atom bomb!

DIAGRAM 24

The basis for the atomic bomb is the splitting of a very heavy and marginally stable shape-resonating nucleus. In the process it, produces enough "debris" to split further nuclei. A chain reaction ensues. The sum total of the mass of the debris and the split nuclei is less than the mass of the original fissionable material. The lost mass is in the form of fast-Zerons, which are released to the surrounding Cosma. These Zerons make the inter-Zeron distance in the Cosma smaller and the frequency of the Cosma increases dramatically. We perceive this as an enormous outpouring of energy in the form of an explosion.

Chapter 22

Cosmatomics

We develop a Zeron based model of Gravity and explain why it is the weakest of the forces of Nature. Gravity is shown to be due to the combination of two radial Zeron fields emanating from massive bodies. We explain why gravity is an instantaneous effect.

At the start of this book we mentioned that gravity and inertia were indistinguishable. The reason will now become clear. Both forces are caused by differential Zeron pressure! If you are starting to get the idea that every force we examine is due to the same effect, you are dead right. Zeron theory is slowly but surely unifying the forces of nature.

Gravity is at once the commonest and least understood forces of nature. As is well known, the attraction between two bodies is directly proportional to their masses and inversely proportional to the square of the distance between them. There is no apparent necessity to involve time in this equation and the conclusion is that the force of gravity is instantaneous in its effect. This would mean that any "signal" consisting of particle or wave would travel at infinite speed, clearly an impossible result. For the last 10 years of his life Einstein sought to incorporate gravity into a Grand Unified Theory that would include the other three basic forces of nature. Ultimately he described gravity in terms of a curved space-time geometry, a concept once again that includes time as a fourth dimension and a variable. It was a desperate last attempt to unify the forces of nature. It didn't make much common sense. It needed an Einstein to understand it.

The reason for the difficulty in understanding gravity and the reason why gravity is by far the weakest of the forces of nature is that gravitational force is not a force, it's the *difference* between two force patterns.

DIAGRAM 25

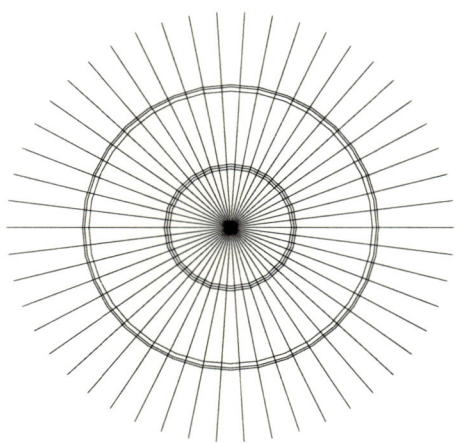

RADIAL ZERON FIELD OF ATOM

To visualize the gravitational process we must go back to re-examine the atom in more detail. The Zeron model of the atom consists of the nucleus and resonant Zeron shells within a Cosma cloud. Up till this point we have represented the Zeron shell and nucleus as discrete entities. In practice there is continuous interchange of Zerons between the atom and the Cosma. The atom is merely a statistical entity. In fact all of matter might be defined as the statistical probability of the presence of an accumulation of slow-Zerons being in a particular place at a particular time. How's that for a strange definition!

Consider the case of an isolated atom surrounded by fast-Zerons of the Cosma cloud. The Zerons move in random directions within the Cosma. The atom absorbs and ejects Zerons continuously. Some Zerons impact the atom normal (perpendicular) to the surface. Others impact at various angles to the surface. These are grazing impacts. However the majority of Zerons that are *ejected* from the atom sensibly fly off normal to the surface of the atom. This is because the concentration of Zerons in either the nucleus or the Zeron electron shell mitigates against a Zeron leaving in any but a normal or near normal

Cosmatomics

trajectory. Each Zeron has, so to say, a multitude of neighbors at the same level, which makes it more difficult to eject in any direction other than normal to the surface.

A massive spherical body produces much the same ejection pattern. Each atom on the surface of the body continues to be impacted from both normal and grazing directions from the surrounding Cosma. Being in relatively close proximity to its neighbors, there is a "shielding" effect one on the other so that the emission of Zerons from the surface of a massive body has a far stronger directional component normal to the surface. There are now two patterns of Zeron resonance surrounding the body. The first is the normal random resonance of the Cosma. The second is an overlay of a radial field. This radial "field" lies at the root of the gravitational force.

DIAGRAM 26

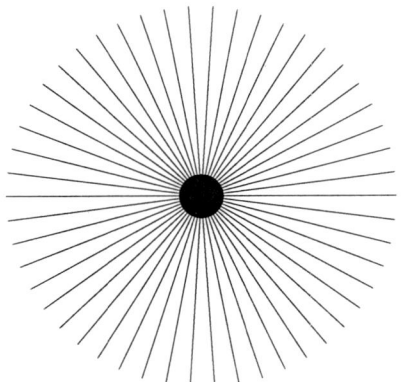

RADIAL GRAVITY FIELD OF MASSIVE BODY

We have defined a field as a non-random pattern of Zeron impact. One such is the gravitational field. This gravitational field can be visualized as straight lines. Surrounding any mass is just such a non-random pattern, which we recognise of impact trains radiating from and normal to the surface of a massive body. This is precisely what is described in classical physics.

Surely this radial field must be a purely local phenomenon? Not so! The ability of the Cosma to transport light beams over millions of light years bears testimony to the loss-free character of the Cosma. The radial Zeron-driven gravity field radiating from any body extends to the edge of the universe.

DIAGRAM.27

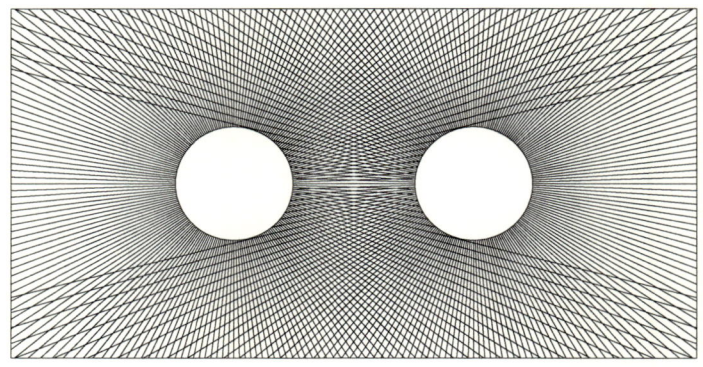

A   B

GRAVITY FIELD OF TWO BODIES

Examining the effects of this gravity field on two masses A and B some distance away from each other, we find that mass A is within the gravitational field of mass B and visa versa. Each atom on the surface of mass A and mass B is bombarded by Zerons in the Cosma producing a Zeron pressure at the surface. On the side of A remote from B, Zerons impact in from all directions ranging at angles ranging from normal to grazing. On the side facing towards B, each point on the surface is bombarded by Zerons from fewer directions because the linear pattern of the gravitational field of B interferes with the true randomness of Zeron impact. The process applies equally to B. In the illustration above darker areas represent a more ordered Cosma less capable of imposing full random Zeron pressure.

The net result of these intersecting fields is a *slightly* greater Zeron pressure on the remote side of each body. The differential Zeron pressure results in a force that pushes the two masses towards each other. This differential force is what we recognise as the force of gravity. Gravity has traditionally been conceptualized as a force of attraction. In truth no such force of attraction exists anywhere within the universe. Attraction is always inverse repulsion. The force pushing any two masses together consists of out-to-in forces from the remote side of each mass.

Being a difference between forces arising from slightly different Zeron pressures and not a force in its own right explains the weakness of the Gravitational force in comparison to the forces which act at the nuclear scale such as the strong nuclear force.

The Zeron theory of gravity also explains the instantaneous effect of the gravitational force. There is no need for a "signal" to alter the force if the inter-body distance is changed. There is no traveling waveform by which the force of gravity is transmitted. Each body is immersed in the field of the other. The overall Zeron-induced gravitational field is effectively static. A movement of one body away from the other is simply a movement of each body into a pre-existent and weaker gravity field. The change in the gravitational force being a purely local effect for each body is therefore instantaneous in its effect.

What about the equivalence of inertia and Gravity? The Cosma is a perfect fluid. Inertia arises from the properties of such a fluid. Gravity arises from the effect of a Zeron field. These are apparently two very disparate origins. The commonality comes from the fact that both forces are due the difference between two Zeron pressures. The forces behave identically.

We stated at the beginning of this book that Zerons have zero gravitational mass. We can see now why this is so. The behaviour of Zerons constitutes the base mechanism for the production of the gravitational force. Zerons as individual entities are not in themselves affected by such fields and thus have Zero gravitational mass.

Chapter 23

The Cosma's Big Bang

We solve the problem of the accelerating expansion of the universe by finding the missing 90% of matter and building a model based on the expansion of gas in a vacuum. All the energy of the universe was injected into the seed that constituted the origins of matter. We find the reason for radioactivity at the outmost reaches of the universe. The ultimate fate of all matter is a return to the hydrogen from which it came with each element becoming radioactive in turn from the heaviest to the lightest. A black hole has a limiting gravity equal to that of the strong nuclear force. Black holes are like giant nuclei and should be detectable by Gamma ray sensitive detectors.

Zeron theory has major impact in the realm of cosmology. It is known from red shift in the spectrum of the light from distant galaxies that the universe is expanding. At the limits of observation the most distant galaxies are receding from us at approximately half the speed of light. Conventional cosmologists theorize that the universe will eventually stop expanding and will re-condense to its original "big bang" state under the influence of gravity. Having theorized this they then calculated that a certain amount of matter would be necessary for the collapse to happen. Then some bright cosmologists calculated the amount of matter in the universe and found that we have detected only ten percent of what is necessary for the reversal of this expansion.

For a decade or more the search has been on for the other 90%. It doesn't take an Einstein to realise that there might just as well be no "dark matter" at all! An equally valid answer would be that the theory that the universe will collapse in on itself is untrue. For some unknown reason that possibility seems not to have been taken seriously. We're going to show that the "collapsing universe" is indeed a faulty model.

We have shown that gravity is the difference between two patterns of Zeron pressure. Pure Zeron pressure is not a differential and is therefore a much greater force than that of Gravity. Even if there were gravity at the start of the universe, it would be a negligible force compared to the forces we are about to discuss. We make mention of this here to pre-empt arguments about the force of Gravity having to be taken into account in the following argument.

Let us imagine the start of the universe. All the Zerons in the universe are packed into a very small sphere. Some clever scientist could calculate the size of this sphere, and I will leave it to him or her. The sphere is dark and quiet. No vibration, no movement, just an immensely dense object, pregnant with untold possibilities.

Suddenly all the energy currently present in the universe is injected into this ball. The Zerons start to vibrate, each one pushing its neighbor away. Of course those Zerons not on the outer skin of the sphere have the same pressure from all sides so they hold their position for a while. The Zerons on the very outer surface of the sphere are pushed from one side only. They launch off into space, releasing as they go the next layer and so on, forming a sort of expanding Zeron "gas". The process is similar to ablation. All this happens because what we know as Newton's laws concerning mass, force and acceleration are already in place. Now we know these laws were installed before the creation! The outward forces caused by the predominantly unidirectional impact of internal Zerons on the outermost Zerons cause these to accelerate outwards forming the Cosma. The newly formed Cosma starts to expand, as would a volume of gas in a vacuum.

Let's play a while with a practical model of gas expansion. Consider a small volume of gas compressed into a rather strong balloon, positioned in the middle of a very large vacuum. If we prick the balloon, the restraint on the gas is removed and the gas starts to expand. The volume gets bigger as every molecule of the gas starts to move away from its neighbor. After a time the gas occupies a larger and larger volume. It will continue to expand for as long as there is a surrounding vacuum. If we examine a single molecule of gas on the outer limit of the gas body, shortly after we prick the balloon, we will see it start to accelerate in a direction away from the geometrical centre of the gas cloud because of the internal pressure of the gas. The pressure is caused by the vibration of, and inter-reaction between the gas molecules. This particular molecule is subjected to a force caused by the impact of adjacent vibrating molecules. As long as there is such a force, this particle will continue to accelerate. The acceleration of this particle will obviously decline as the internal pressure reduces and the corresponding imposed force on the particle reduces. However, the particle will continue to accelerate for as long as there is an applied force. Therefore for as long as there is any residual internal pressure, a molecule on the outer fringe of the gas cloud will continue to accelerate away from the center albeit at an ever decreasing rate of acceleration. The same applies to every molecule in the gas cloud. This is precisely the description of the Zeron model of the universe.

It has very recently been discovered that the outer reaches of the universe are not only moving away from us at large velocities, but are accelerating away from us. Conventional cosmologists now recognise that there must be some as-yet-undiscovered force overcoming gravitational attraction. There are no suggestions as to what that force might be. Now we know.

If the origin of the universe was an extremely compact body of Zerons (the seed Cosma) was suddenly injected with energy and released into the vacuum of space, the Cosma would expand precisely in the manner as described above in the gas model. The outer reaches of the Cosma would be accelerating away from us, as would every particle of matter between us and the edge of the universe. It will continue to do so indefinitely. All the above neatly solves the problem of the "unknown force" causing the universe to expand at an ever increasing rate!

What are the implications of the expanding universe for us on earth? Our whole environment expands in concert with the universe. Even our bodies expand but that is not as bad as it sounds for slimmers. This expansion effect is of course not measurable as everything around us is subject to the same expansion. There is no experiment you could perform to measure this expansion. All your instruments and apparatus will be subject to the same rate of expansion. Ironically there *is* a natural process that confirms this expansion. It's called radioactivity! See the next sedation for this rather bizarre but totally credible theory.

What lies beyond the Cosma cloud? The fact that the universe appears to be expanding and is even undergoing accelerated expansion tends to indicate an unfettered process. There is a good chance that beyond the universe is God's Infinite Nothingness.

Radioactivity

Radioactivity has been in the realm of "popular physics" since the days of the Curies. The lethal nature of its energy outbursts was not initially recognized. Madam Curie died of radioactive poisoning. She actually carried samples of radium around with her while they bombarded her body with the unseen but deadly bullets! Radioactivity has been widely used in the medical, industrial and scientific fields ever since that time and yet it remains one of the mysteries of modern Physics. Why *should* certain elements break down spontaneously and what is the mechanism of the breakdown? Many volumes have been written on the subject and indeed a prime force of nature has been conjured up to explain the phenomenon, - the weak nuclear force. According to CP the weak nuclear force is responsible for radioactivity and is described as an electromagnetic force. Once again one force is described in terms of another force which itself has no reference to real fundamentals. It's a very weak force and a very weak explanation. The Zeron theory of radiation is extremely simple. Astonishingly, the root cause of radioactivity is not to be found in the atoms from which it emanates, but in what is happening at the outermost reaches of the universe!

We saw in the previous section that the universe is expanding at an ever-increasing rate. Because there is a fixed and finite number of Zerons in the universe*, the Zeron density in the Cosma cloud is always decreasing albeit at an undetectable rate. (*This presupposes that the conversion of matter into energy and visa versa is constant within the universe. Evidence from our astronomers indicates that this appears to be the case.)

The radioactive atom is a very wobbly thing. It shape distorts into elongated and flattened shapes. In the elongated shape it is very vulnerable to disintegration. If the "neck", that forms when the nucleus elongates, gets even minutely too small, it never stops narrowing. The atom snaps into bits and Zerons are ejected until the two smaller bits achieve stability by resonating. We perceive the scattering of these Zerons as the emission of energy. This is the energy of radioactivity. A lump of radium has atoms at all levels of excitement, some on the very edge of breaking apart. Others are more stable. (There's a neat quantum theory that correctly deals with why this is so. We won't bore you with the details.) If the Zeron pressure that holds the nucleus of the atoms together reduces, some of the atoms that were "on the edge" will distort past their limits and will disintegrate. This is precisely what is happening to cause radioactivity in certain materials. The Cosma expands. The Zeron population within the Cosma becomes less dense. Zeron pressure on the nuclei of the atoms reduces, and atoms that have been shape changing to the edge of disintegration, now disintegrate. This "spontaneous" decomposition takes place sporadically, but taken on average, it takes place with clock-like precision. This precision comes from the "steady" rate of expansion of the universe! This is why the half-lives of radioactive materials are measurable and constant.

Of course we know from the preceding paragraphs that "steady" is not quite correct. The expansion of the universe and consequently the rate of decomposition in radioactivity are not quite constant. The expansion of the universe is accelerating and so therefore is the rate of radioactive decay. Don't expect to be able to detect the changes in radioactivity due to the expansion of the Cosma or to the accelerating rate of expansion. Our clocks, even atomic ones, are entirely dependent on Zeron density and are regulated by exactly the same expansion phenomenon making the change undetectable.

Can the process of radioactive decay be altered or halted? Remember we are dealing with a Cosma that pervades all space and all matter. To try and build a barrier to isolate a "radioactive" material from the influence of the universe would be like trying to build a waterproof tank out of one-inch mesh. It simply can't be done. Radioactivity faithfully obeys the exponential decay formula typical of such processes regardless of where the material happens to be even if that is just this side of nothingness.

What happens if the expansion of the universe continues indefinitely? Well the lady of two billion and two will not be wearing golden jewelry. Gold will be highly radioactive! As Zeron pressure reduces, the radioactive process will move down the food chain, with the Cosma finding ever lighter materials to disintegrate. Some atoms are inherently more stable than other of similar atomic weight because of their nuclear configuration, so there will be a delay while the Cosma feeds on equally heavy but less stable cousins. But even the stable ones will be attacked in due course. The process is inexorable. Eventually all will be reduced to Hydrogen from whence it came.

Superneucleonic Black Holes

The only circumstance in which Zeron pressure is completely from outside-to-inside is as occurs at the hard core of the atomic nucleus. There are no fast-Zerons within this core. It follows therefore that the limit of such Zeron pressure may occur when a massive body is so condensed by gravity that it is like a huge nucleus. This super-nucleus would constitute the core of the "black hole". The limiting gravitational force can be found by comparing the gravitational force to the "strong nuclear" force. The strong nuclear force is a factor of 10^{38} times the gravitational force. Thus a black hole will have at its surface a gravity of 10^{38} G! This is the limiting value of Zeron pressure and therefore the limiting value for gravity.

Strictly speaking, light cannot leave a black hole, but not as is conjectured because the extremely high gravity field prevents photons from leaving the body. It is because the core material of a super-nucleus has no electron or nucleonic shells. Without such shells there is simply no mechanism for the generation of radiation. Absolutely no radiation is formed in the hard core of an atom or in the hard core of a black hole. However, as with the atomic nucleus, there is a mega halo of decreasing density around the "pure" black hole core. This halo has resonance shells, which can generate Gamma radiation just as in the atomic nucleus. There will certainly be enough stimulation of the black hole shells from in-rushing matter, so it is here that gamma radiation will be generated. Black holes should be detectable by this radiation probably in the high x-ray or gamma ray wavelengths. Perhaps this is the origin of the unexplainable high-energy x-ray emissions that we spoke about in the opening paragraphs of this book. The new Gamma ray detector telescope in currently being constructed in South Africa will pinpoint these sources of radiation. Without Zeron knowledge, will they know what they are looking at?

Chapter 24

Homeopathic Memory

We examine the mystery of homeopathic medicine and how it can be that homeopathic solutions containing none of the original dissolved substance still work as if the solution is at original strength or better. We re-examine well publicized trials of homeopathic solutions, and develop a new model analogous to permanent magnetism. Water clustering is revealed as the mechanism behind homeopathic memory.

From the sublime of the infinite cosmos we revert to the almost ridiculous phenomenon of homeopathic medicine. Homeopathy is based on the premise that diseases that cause certain symptoms can be cured by substances that produce the same symptoms. The second premise is that the more dilute the homeopathic medicine, the more potent it is. We will be addressing the second of these premises.

The substances are given to the patient in ultra small doses. This is accomplished by putting one drop of the substance in 100 drops of water. This is called C1. One drop of the resulting solution is then mixed with a further 100 drops and this is called a C2 solution. C30 solutions are commonly used. If we calculate the dilution ratios of a C30 solution we find that it contains $10^{2 \times 30}$ parts water to one part substance i.e. one part substance in 10^{60} water. Now this is an extremely large number. It is 10 followed by 59 zeros.

To put this into perspective a C12 solution would be one drop of substance in 20 swimming pools of water. A C15 solution would be one drop in the Atlantic Ocean. A C20 solution would be one drop in all the oceans of the world and there would not be enough water on earth to represent the one-drop of substance in a C30 solution. To put it another way there is about one chance in a billion that a drop of C30 would contain even one molecule of the substance. Clearly there is virtually no chance that the substance itself has any effect on the patient. And yet homeopathic medicines seem to work! How can this be? Perhaps the success is due to the well-known placebo effect where a relatively high proportion of patients respond positively even if they are given sugar pills? However homeopathic medicines are routinely used by some vets and it is most unlikely that animals would be subject to a placebo effect.

In a television documentary flighted on television by Horizon in 2002, the story is told of two reputable scientists who put their very solid reputations on the line by confirming that, according to their experiments, the homeopathic water of a C30 solution actually worked and the only possible explanation was that the water had developed a memory. Under the auspices of two highly prestigious organizations, Nature Journal with Sir John Maddox and The Royal Society with Prof. John Enderby, sophisticated triple-blind tests were performed. These tests showed that the homeopathic water had not developed a memory and that it had no therapeutic power. What is going on here?

The initial problem in getting the conventional scientist to believe the results of carefully conducted result of experiments by other reputable scientists is that there is absolutely no conventional scientific explanation of how water could develop a memory. Two meticulously performed experiments showed that there was no memory and those who performed the experiments did not see a necessity to backtrack to see if there were any holes in their experiments. The results agreed with their preconceived ideas and therefore must be right.

Back to Zeron Theory. If water or other solvents have a memory it is because there is something going on at the Zeron level. It is known that homeopathic solutions can be stored indefinitely without losing their efficacy provided they are isolated from heat and light. This sounds like the Zeron environment where every vibration once set in motion continues forever without diminution. If we consider the hydrogen atom we would find that there is a fundamental frequency of the nucleus determined by the resonance of its Zeron components and the frequency of the surrounding Cosma. As the nucleus becomes more complex more complex vibrational signatures are set up. For every substance there is a unique spectrum of frequencies. We believe that it is this spectrum that lies at the root of the Homeopathic phenomenon.

Close examination of the Horizon video shows homeopathic dilutions being prepared both in the conventional way and by the two laboratories that conducted the rigorously controlled tests. In the conventional procedure, one drop of a substance is added to 100 drops of water and the tube is shaken and then banged a number of times on the table. This procedure called sucussion was stated to be essential in the production of Homeopathic solutions. This procedure is repeated up to 30 times to produce what is known as a C30 homeopathic solution.

In the laboratory process one drop of the substance is added to 100 drops of water and the test tube is inserted into a rotating head of a mixer which gives it a good shake. This appears to mix it thoroughly. At first glance there seems to be no difference in the two procedures. Closer examination of the two procedures seems to reveal that in the lab procedures a circular motion is imparted to the water whereas in the conventional procedure, sharp shockwaves are transmitted to the water in the process of banging the test-tube on the table. Even if the mechanical mixing apparatus was used in both cases, it would depend on the operator's technique as to whether the test-tube was subjected to impact or merely shaken in a circular fashion. Surely impact cannot make much difference? We think it does.

To use an analogy, consider a permanent magnet. Magnetism occurs when the "spin" in the atoms are aligned in the same direction. At subatomic level Zeron theory tells us that we are dealing with a totally lossless and therefore undamped system. This is borne out by the fact that once a permanent magnet has been magnetized, it stays that way forever. The magnetic material has developed a permanent memory! How does one make a permanent magnet? We place the un-magnetized bar in a magnetic field and tap it repeatedly to produce shock waves in the material of the bar. This makes the atoms line up in one direction and miraculously the bar has become a permanent magnet. To destroy the magnet one can either heat it to disrupt the pattern of molecules or one simply hammers it into submission.

When the initial homeopathic solution is mixed, the substance's molecules get dispersed into the water in the ratio of one in a hundred. Now we return to Zeron theory.

Solution takes place when the mode of vibration of the substance's molecules can be imitated by the molecules of the solvent. If the water molecules cannot duplicate the vibrational mode of the substance, then the substance is insoluble in water. The same principle applies to all other solvents. When the homeopathic substance is dispersed into water, the water mimics the vibrational mode of the substance and the substance dissolves.

The peculiar part about this phenomenon is that the whole volume of water appears to adopt the same vibrational mode (resonance effect), so that in spite of the extreme dilution of the substance, the water behaves as if it is at the original concentration or even at an increased concentration. No matter how often the water is diluted, if the solution is subjected to adequate shock to provide the energy to induce the changes, the water remains minimally at the original dilution as far as its efficacy is concerned. At higher C numbers appreciably higher efficacy is observed. Only the substance itself gets diluted which in the case of toxic substances can only be a good thing. The other peculiar thing is that this "memory" is, like the magnet, a *permanent* phenomenon unless steps are taken to destroy this memory. It is suggested that the memory effect can be experimentally determined if a C30 solution is split into two batches and one batch is heated to just below boiling point. The prediction is that the efficacy of the heated solution would be zero as compared to the efficacy of the unheated sample.

Once more, we have presented the simplest scenario. In practice the detail is far more complex. When we talk about water being able to mimic the vibrational spectrum of the substance dissolved in it we are talking about very complex vibrations. To draw an analogy, consider the spectrum of sounds emanating from a full orchestra of instruments all playing the same note. An analysis of the output sound would show a different intensity of sound for each frequency band. The entire picture would represent the sound spectrum for that particular orchestra playing that note. If one instrument plays a different note, the spectrum changes. There is obviously an infinite number of spectra that could be produced.

Similarly every substance has its own peculiar spectrum. How do we know this? Modern nuclear resonance apparatus is easily able to detect this spectrum. Experiments on solutions of substances and their C30 homeopathic equivalents (which contain virtually none of the original substance) show identical spectrums to that of the original solution.

How does the relatively simple water (or other solvent) molecule mimic what might be an extremely complex spectrum? One model might be based on the vibrations of the atoms. Firstly, for every atom there is whole range of potential frequencies. In each nucleus we found that there were resonance shells equivalent to the electron shells. For each such shell there is a resonant frequency in the gamma and x-ray range. Then there is the lower resonant frequency of each electron shell and the still lower frequencies of the quantum shells. Then there are the "spin" frequencies which are caused by the continual deformation of the atom. These deformations can be very complex, and produce a whole spectrum of frequencies on their own. Finally at molecular level the combination of all these frequencies produce and interactive "beat" frequencies caused by constructive reinforcement and destructive interference of their interactions. These characteristics alone would provide a huge spectrum of frequencies that might enable the solvent to mimic the spectrum of a soluble substance. However it is believed that to alter any of these fundamental frequencies would be to change the very nature of the solvent molecule and this is an unlikely scenario.

There is one other phenomenon which would enable a solvent to mimic spectra. It's the phenomenon of water clustering. Homeopathic water consists of large clusters of the water molecule as distinct from the individual molecules of distilled water. This can be demonstrated by freezing distilled water and homeopathic water. The ice crystal formation of any homeopathic solution is distinctly different from that of distilled water. Furthermore magnetic resonance measurements show a distinct size distribution of the clusters within any solution. As the C-number for the homeopathic solution increases the size distribution curve narrows and an ever larger number of clusters conform to a mean size. This probably explains why the higher homeopathic number solutions are more potent. It is therefore very likely that the water cluster model is the valid one as there would be almost limitless ways in which large clusters of atoms could be induced to mimic the resonance patterns of solutes. The same principle applies to all solvents

Yes solvents do have a memory. They adopt the vibrational spectrum of the substances dissolved in them and the effect is permanent providing nothing is done to destroy that memory. In the light of the above we believe that there is a good case for the experiments to be repeated. This would advance the cause of physics and also help those whose reputations were damaged by the apparent failure of the carefully controlled experiments.

Chapter 25

It Happens Every Day

Differential Zeron pressure lies at the root of many of the common physical phenomena that are part of our everyday lives. Any differential in fast-Zeron pressure on a body constitutes a force on that body. Fast-Zeron activity within solids, liquids or gases is always less than in a vacuum. Likewise activity within solids is less than in liquid which is in turn less than in gases. It is this fast-Zeron activity that lies at the root of all the everyday physical phenomena that surround us. A few examples are presented here.

Solids Liquids Gases and Plasma

Have you ever read a good explanation of the reason for the existence of the different phases of matter? I haven't. Zeron theory provides a simple model for these three states of matter. With some exceptions most materials are solid at relatively low temperatures, liquid at intermediate temperatures and gases at high temperatures. For any particular material, at the lower temperatures, when the material is solid, fast-Zeron activity within the material is at a relatively low level. This is because "temperature" is a function of and is a manifestation of fast-Zeron resonance frequency. Fast-Zeron activity is present within all materials. Only within the core of the nucleus of an atom is this fast-Zeron activity absent. At relatively higher temperatures when the material becomes liquid, the Zeron resonance frequency is higher and likewise is higher still when the material gasifies. In solid form, the difference between internal and external Zeron pressure results in a differential "out-to-in" Zeron pressure. This net pressure is sufficient to interlock the atoms that are then able to resist shear forces. In the liquid phase, this differential pressure is less and the resistance to shear reduces suddenly to a low level. The interlock effect is largely lost but the atoms still hold together under the influence of the differential pressure. In the gaseous form, the differential pressure is insufficient to overcome the natural repulsive forces between electron shells and the disassociated atoms repel each other to form a gas. The gas will expand indefinitely unless restrained. There is one other form of matter. It is called Plasma. Plasma according to conventional physics is a gas from which all the electrons have been stripped. Plasmas are formed at very high temperatures. Zeron theory has a different view. Its not that atoms are stripped from the plasma, but that at very high temperatures, the frequency in the Cosma is such that the atom's nucleus simply cannot keep up to form a resonating interaction with the Cosma. No electron shells are formed at all. Once more Zeron theory presents simple and logical explanations.

Tensile Strength

We can get some idea of the enormous forces imposed by Zeron pressure if we look at the way materials develop resistance to being pulled apart. The key to all tensile capability in materials is the fact that there is less fast-Zeron activity inside solids than there is outside them. In the same way that differential Zeron pressure acts on the nucleus of the atom to hold it together, so on the macro scale differential fast-Zeron pressure holds materials together by compression. The difference between internal Zeron pressure and external Zeron pressure constitutes an external-to-internal force, which resists any tensile force imposed on the object. We get some idea of the magnitude of this Zeron pressure if we try and pull a small diameter steel bar apart. One needs a very powerful machine to do this. It's incredible to realise that this very high tensile force is developed by a *differential* Zeron pressure acting on the ends of the rod. The magnitude of the force is moderated by the crystalline structure in the material. It seems clear that there is more fast-Zeron activity in the planes between crystals than within them. It is for this reason that single crystals of substances exhibit much larger tensile strengths than their crystalline matrix counterparts. Simple isn't it?

Surface Tension

Water or any other liquid sprayed into air (or vacuum) forms spherical droplets. Differential Zeron pressure results in a force directed towards the centre of mass of the droplet, causing it to assume minimum volume, which is a sphere. It's fascinating to watch the astronauts playing around with free-floating globules of water in the Shuttle. It is also instructive to see exactly how the globule of water changes shape. This is exactly what happens to the nucleus of an atom within the Cosma cloud. Differential Zeron pressure is responsible for all surface tension effects.

Dust Adherence

Housewives can blame the Zeron for the curse of dust that adheres to everything in sight. Small solid particles (dust or powder) will adhere tenaciously to any surface. That this is not an "electrostatic" phenomenon can be deduced from the fact that dust or powder will readily adhere to an earthed metal object.

Differential Zeron pressure lies at the root of this phenomenon. The dust particle comes into contact with a surface and settles, making contact with the surface at three points. At the points of contact, fast-Zeron activity is inhibited compared to the Zeron activity on the rest of the dust particle. This leads to a net force that pushes the dust particles towards the surface. We interpret this as adherence. The only way to get the dust particle to release its grip is to apply a greater force than the force of adherence by using a duster or cloth. Having removed the dust particle, it gleefully clings to the cloth by the same means! You never thought that such an everyday phenomenon had an answer in advanced physics!

Glue Solder Brazing Welding

All these joining techniques depend on the formation of a solid between the two bodies that are required to be joined. Fast-Zeron activity is less within the joining material than it is outside the bodies being joined. Differential Zeron pressure provides the compressive force that holds the bodies together. The degree of adherence depends on the extent to which the "glue" excludes fast-Zeron activity.

An interesting effect that tends to confirm this finding is found with engineering gauge blocks. These are blocks of metal that are machined to very high tolerances and have, as far as can be attained, perfectly flat faces. If one slides two blocks together, so excluding air from between the surfaces, it is found that the blocks cling together tenaciously. The force required to part them far exceeds the force to break the vacuum between the blocks. This effect has never been explained. It's simply that by mating two accurately machined surfaces together we inhibit fast-Zeron activity between them and the external compression comes into effect. It's a sort of glue-less glue.

String Theory?

No this is not about the latest scientific theory proposed to solve the mysteries of the universe! It simply asks the question: How does a piece of string work? The answer may seem surprising at first but a little reflection will show that the Zeron interpretation of even a simple example such as this makes eminent good sense.

Place a piece of string on a table and take one end in each hand. You will not be surprised to find out that if there is slack in the string there is no resistance against any tension you try to apply. This is because the only component of Zeron pressure on the string that can resist a tensile pull is along the length of the string. Because this component is nowhere aligned and because any pull on the string initially realigns these components without achieving much else, there is no resistance to the pull. Pull further and all these "along the length" components of Zeron pressure line up exactly and act in concert. In effect what one is left with is a resultant inward force at each end of the straight part of string resisting your pull.

Don't believe me? Well if you now cut a piece of string from the middle with nice neat right-angle cuts and clamp each end neatly so that the Zeron pressure can act on a nice plane face, what do you get? Exactly the same result! The fact that the ends of the string may be wound around your fingers does not influence it at all. The Zerons are very clever at applying pressure to any configuration of material, even if it's a piece of string wound round a finger.

Through the Looking Glass

Clear glass is transparent - right? Wrong! When the light wave hits the surface of glass it is absorbed, transformed into Zeron energy, re-emitted as light energy, and so on millions of times, and eventually transformed into light and emitted on the far side of the glass! And Zeron theory was supposed to make things simpler?

Let us look at the steps listed above. Firstly, provided the light ray is composed of the correct range of frequencies, the random impact of the Zerons on the surface of the glass, (which is what holds it together), is modified by a train of lateral vibrations representing the incoming light wave. The molecules on the surface of the glass absorb the light wave packets, get excited momentarily, the electron shell takes a jump, finds itself out of synch with its environment, and reverts back to a more comfortable status, emitting an identical wave packet. This gets passed on from molecule to molecule with each performing in an identical way. Ultimately the molecule on the far surface of the glass behaves identically to its corresponding molecule on the near surface and emits essentially the same wave packet as was received on the near side. Within the glass there is essentially a complicated absorption and emission sequence that transfers energy from molecule to molecule rather than a direct transmittal of light energy through the glass. Now this process takes time and it is therefore not surprising that the passage of light energy through glass is accomplished at a lower velocity than the speed of light through a vacuum. Now glass is a nice amorphous type of material, (actually it is a liquid) with no internal crystalline faces to mess up the undistorted transmission of the energy waves. And so at the other face of the glass, the surface molecules are forced into a carbon copy behavior of the atoms on the near side. Provided the far side of the glass is plane and parallel to the near side, the pattern transmitted is identical to the pattern absorbed. We perceive the glass to be "transparent"! If however the far side of the glass is not parallel to the near side, the emission of the light packets is direction influenced in accordance with the plane of the far side. This is of course the way we make prisms and lenses.

If the wavelengths are such that the vibrational frequency of the incoming wave is not close enough to the natural frequency of vibration of the molecule, no absorption of the incoming energy takes place and the glass is perceived as being opaque to those frequencies. Glass, for instance, is a poor transmitter of infrared and certain ranges of ultraviolet light. It can be deduced therefore that light transmittal through glass is essentially a resonant phenomenon.

What's the solution?

Put salt into water and agitate it. The salt disappears and you have what is called a salt solution. Exactly what is a solution? Nobody is saying. Zeron theory provides a fundamental answer. In the chapter "Homeopathic memory" we described the phenomenon water clustering. Now water as we know is an excellent solvent. It is so because it is able to form large water molecule clusters, which in turn are able to mimic the vibrational spectrum of many substances. If it can so mimic, the substance dissolves in the water. If not, the substance is insoluble. The deformation patterns within water are not confined to a few molecules only. The entire body of water is subjected to a resonance effect which ensures that every molecule of water behaves in an identical fashion. This accounts for the phenomenon that even in extremely dilute solutions, some molecules of the dissolved solute will be found in equal proportions in every cc of solution.

Chapter 26

Is the Cosma Real? The Clincher

The "undetectable" Cosma comes out of obscurity for a few seconds each day in the form of Gamma Radiation Bursts! Long Gamma Bursts have been identified as coming from cosmic events. Short Gamma Bursts constitute the greatest Mystery of modern astronomy. We identify Gamma bursts as being the spontaneous emission of Zerons from the expanding Cosma. Gamma ray bursts have been identified as coming from the surface of the earth. One Bizarre evidence for Gamma bursts on the earth's surface is the incidence about once a year of human spontaneous combustion. These have all the hallmarks of such a burst. We show the possible link to between gamma ray bursts and cold fusion experiments.

The great Gamma Burst mystery

One of the major problems with Zeron theory is that we are dealing with what is essentially an undetectable form of matter. In conventional physics, if you can't detect it, it doesn't exist. At most, the Zeron would be designated as a "virtual" particle. A virtual particle is a bit like the hole in a donut. It is there but it consists of nothing. In spite of the overwhelming circumstantial evidence of the existence of the Zeron and the Cosma presented in this book, Zeron theory is always subject to doubt and criticism because of this type of conventional scientific methodology and thinking. Well, we have saved the best for last, because (surprise, surprise) the elusive Cosma pops up out of apparent nothingness to manifest itself in the detectable world for a brief few seconds at least once every earth day! So we don't actually have to detect it to know its there. All that would be needed to actually *see* it during these brief manifestations would be Gamma-ray-sensitive eyes.

It has been known for some time that about twice a day on average, the universe is lit up by highly energetic bursts of Gamma ray energy. If our eyes were sensitive to Gamma radiation we would see an intensely bright spot in the sky which, like the sun, would light up the entire sky for between one and two seconds, making the full light of day seem insignificantly dim.

There are two different types of Gamma Bursts. One is short and hard (one to two seconds duration and of higher energy) and the other is long and soft (minutes in duration and of much lower energy). The long soft emissions are relatively few in number and one or two have been positively linked to optically observable cataclysmic cosmic events in remote galaxies.

The origin of the much more numerous short hard variety is unknown. Short gamma ray bursts have been called the greatest mystery of modern astronomy. They cannot originate from cataclysmic happenings in space for a number of reasons. Firstly, the emissions are basically non-directional. The source points of the energy are detected in every direction. The distribution is entirely random and evenly distributed and the radiation has never been observed to come from the same direction twice. If these bursts were to originate from what is a non-isotropic distribution of galaxies in space, then the distribution of such bursts would also exhibit some "lumpiness", but they don't. Secondly, if the observed short gamma ray bursts originate from distant galaxies the total energy of *each* of these bursts would exceed the energy emission of the entire mass of the universe converted to energy in a few minutes. These characteristics rule out remote events as a probable source of these bursts. The cosmologists respond by saying that the event that produces such incredible outbursts of energy are detectable only because they "beam" the light out in specific directions and we happen to be in the searchlight's beam. It's a false argument.

Assume that there is a phenomenon that enables such beaming to take place. If the beams were to have an emission angle of 1 degree, then in any one plane there could be 360 directional possibilities for the beam. If then one were to rotate that plane by one degree at a time, there would be 360 such planes making a total of 360 x 360 possible directions for the mission to head off in. This would mean that there were 129600 occurrences of the phenomenon for every one we see. A more likely scenario is that the angle of emission would have to be far, far smaller, probably a thousand times smaller. Now we get into really big numbers because 360 000 x 360 000 = 129 600 000 000 actual occurrences for every one we see! In fact the theory of beamed emissions does not work because the math says that the more tightly the beam is focused, the more are the occurrences that we don't actually see and the mass that needs to be converted to enable us to see all those emissions remains sensibly the same. We just don't ever convert that amount of mass into energy.

Of course if the emissions originate in our immediate vicinity, these energy outbursts could be very modest indeed. It is only on the assumption that they originate at great distances from us that we interpret that they need to be extremely powerful in order to reach us at the observed intensity. Just to add to the problem of exactly where these bursts come from, it was reported a satellite-borne gamma ray detector had detected some of these bursts as coming from the surface of the earth.

Short gamma ray bursts constitute an unresolved 30-year-old puzzle. Scientists still have absolutely no idea about their origin or mode of formation.

There's an easy answer, and to find it we need to return for a moment to the mechanism of Gamma ray production as discussed in Chapter 10. There we found that Gamma ray emission resulted from the collapse of the outermost Zeron shell of the nucleus of the atom to a smaller radius. In effect the Zerons in this shell have changed resonant frequency, and the excess Zerons and therefore energy are emitted in the form of Gamma radiation. Gamma radiation is therefore the result of a change in the resonant frequency of Zerons.

Now we know that the universe is expanding. The Cosma, by definition, is also expanding and the mean distance between Zerons in free space is gradually getting larger and larger. Because the velocity of free Zerons is a constant, there is an unstoppable tendency for the frequency of vibration of the Cosma's Zeron population to reduce. However, the Cosma is a resonating entity and it is not free to assume a random frequency. It looks for the next lowest *resonant* frequency. As the inter-Zeron distance increases the Cosma tries valiantly to maintain its current resonance. It can only do so by forcing the Zerons to do something highly a-typical. The velocity of the free Zerons increases marginally to maintain resonance frequency. Faster Zerons mean higher energy and the energy level in the Cosma increases. This cannot be sustained indefinitely and eventually the Cosma lets go at a weak point. The frequency collapses to the next permissible number and the excess energy is released in the form of Gamma radiation. The Gamma radiation sweeps outwards at the speed of light and the initiation point or seed point is detected as being the origin of the Gamma burst.

Seed points are formed when the combination of all Zeron wave patterns due to radiation, gravity and the existence of a quantum field momentarily produce a super-high Zeron velocity at a particular location in the Cosma. The Zerons within this seed location are the first to assume the new resonant frequency by reverting to normal velocity. This new frequency spreads in every direction, producing Gamma radiation as it goes. After the collapse, the velocity of the Zerons returns to normal, and the entire cycle starts over again.

The short bursts of energy that our Gamma ray instruments detect don't actually come from anywhere. We are part of and are immersed in the Gamma radiation engine itself. The fact that these bursts seem to come from a particular but random direction can be accounted for by the fact that the burst will appear to emanate from the seed point. However every cubic centimeter of the Cosma makes its own contribution to the burst. One of the characteristics of this radiation is that no amount of shielding seems to attenuate the burst. This could only be the case if the shield itself is part of the Gamma engine. This is further borne out by the satellite detection of gamma ray bursts emanating from the earth's surface. The earth too is part of the Cosma and these bursts occurred because the seed point was in the earth or near the surface of the earth.

If there was ever any doubt about this elusive Cosma, here is the real clincher to prove that it exists. The only possible source of such isotropic energy is space itself. There is no other vaguely feasible explanation for these Gamma ray outbursts other than this one, based on the properties and behaviour of a resonant Cosma. In effect, the Cosma jumps out of obscurity into the observable world for a few seconds every day of our lives! For these few seconds the Cosma and its component Zerons are real and detectable. Then they jump back into apparent nothingness. No longer can the Cosma be called a "virtual" system or the Zeron be a "virtual" particle. They are as real as you and me.

Is the Cosma Real? The Clincher

Bizarre possibilities

Gamma ray bursts may provide some answers to other puzzles, some quite bizarre. How many forest fires devastate parts of the earth every year? Every year sees huge forest or bush fires on most continents. The cause of the fire is always accredited to arsonists or to lightning. Strangely, most fires occur in drought times when there is no lightning. Could there really be that many arsonists? Then again, each year, many building are razed to the ground by fire. Forensic investigation can usually pinpoint cases of arson, but the majority are put down to "electrical faults". I wonder!

Then there is the problem of strange lights in the sky, usually interpreted as being manifestations of UFO's. Alternative explanations include flares, balloons, military experiments etc., but there genuinely seems to be some mysterious force at work. Gamma ray bursts?

There is one bizarre but well documented occurrence for which there is *no* alternative explanation, - spontaneous human combustion! There are many well documented cases of people simply combusting for no apparent reason, leaving behind a small heap of ashes. These events happen about once a year worldwide, and there is much photographic evidence of the gruesome but highly unusual burn characteristics. A foot here or a hand there is left on the edge of the ash pile, completely unscathed. Even newspapers lying right next to the ashes remain un-scorched. Very high, very short duration temperatures would be required to completely incinerate a body (including the skeleton) and leave nearby materials untouched. There is no known physical process that could produce such results.

Could gamma ray bursts be at the root of all these phenomena? It is clear that at the seed point of gamma bursts there would be an intense generation of energy, mainly because the gamma ray is an exceedingly energetic form of radiation and because at the seed point, that energy would be spatially concentrated. It's like throwing a stone into a still pond. A short distance from the "plop", a low intensity wave front proceeds outwards causing little turbulence as it passes by. At the point of impact, however, there is a much higher energy concentration as the energy of the stone gets absorbed into the water. At the seed point of a gamma ray burst, there could be countless billions of Zerons all simultaneously breaking the speed limit. When the Cosma "gives" all this energy is released instantaneously in what could be a very small volume. Anything within this volume would be subjected to intense radiation energy. In conventional terms, there is no known way to transform gamma radiation into heat, but in Zeron terms, heat production would be the inevitable result of a gamma outburst due to undue resonance in the Cosma.

If, as in a forest, there is enough combustible material both inside and outside the gamma burst seed point, a forest fire ensues. If the seed point were in the atmosphere the air would become superheated (as occurs in a lightning bolt), causing the mysterious lights to be seen at that point. In the case of some UFO sightings, it was reported that the electrical systems of motorcars in which the observers were traveling simply stopped working. If one looks back at the Zeron theory of the electron it becomes evident that it is entirely feasible that a change in the Cosma's resonant frequency might well disrupt the generation and flow of electricity, at least until the whole system is back in step.

Perhaps a more scientific experiment showing evidence of gamma bursts is the so-called "cold fusion" experiments conducted by Fleischmann and Pons. They applied a constant current through a cell containing a palladium electrode in heavy water. The results have been shown to be highly variable and most times unrepeatable, and as a consequence there has been a great deal of skepticism and disbelief of what they have reported. The most spectacular result they reported was that in one cell the most of the electrode melted and part of it vaporized, destroying the cell and the fume hood enclosing it. We believe that their experiment edged a localized volume of space closer to the gamma burst conditions in the Cosma. When resonance occurred there was a high outburst of energy. We believe what Fleischmann and Pons reported. However, repeatability is in the lap of the Cosma.

This all seems to make reasonably good sense. The real decider however is the evidence of those gamma ray bursts detected on the earth's surface. Gamma ray bursts *do* occur at ground level and this lends credibility to the models created to explain the above phenomena. It therefore seems entirely possible that these short gamma bursts jump out of the Cosma, generate very high temperatures in small pockets, incinerate everything within those pockets, make bright lights in the atmosphere, create various other strange phenomena, and then disappear without trace back into the Cosma.

The Cosmic Microwave Background

If the above sounds a bit far-fetched, there is a far more conventional evidence for the existence of the Cosma. It is the so-called Cosmic Microwave Background or CMB. This is supposedly the leftover energy from the Big bang. There are two unsolved mysteries regarding this microwave remnant. The first is that the energy is remarkably the same wherever you measure it. In any direction this energy is the same within one part in 100000. This extraordinary consistency can only mean one thing; it originated at the Big Bang. The reason for this is that even at the speed of light there is no way the extremities of the universe could keep in touch to maintain this evenness. It must have been there from the beginning.

Now the CMB has got one very significant property. If one looks at the frequency of the CMB it is found that it has a range of frequencies in the microwave spectrum. If one plots this spectrum it is found that the strength of this radiation for each frequency results in a familiar skewed bell shape. It's Planck's blackbody radiation curve!

Now we previously discovered that Planck's Constant turned out to be the energy of a fast Zeron. It follows that the curve that it relates to somehow represents the properties of the Cosma which is composed of these fast Zerons. Now suddenly all the pieces of the puzzle fit together. The Cosmic Microwave Background is indeed a manifestation of the Cosma.

There's more. By measuring the red and blue shift of this radiation in different directions it has been found that our group of galaxies is traveling through this radiation background at a speed of more than 600 km per second. Now we have the result that not only is the Cosma the universal frame of reference, but that we are traveling through it at a known velocity. Of course we still can't detect the Cosma directly or perform a Michaelson and Morley type experiment to measure anything at all, but it's nice to have confirmation that the Cosma has an indicator like the CMB. One commentator puts it this way:

"By studying the acoustic signals in the CMB, cosmologists have estimated the age, composition and geometry of the universe. But the results suggest that the biggest component of the modern cosmos is a mysterious entity called dark energy".

Indeed the entity otherwise described as the missing matter of the universe is our old friend, the Cosma.

Chapter 27

The End in Sight

We conclude by summarizing what we have discovered from Zeron Theory. There is a Cosma consisting of fast Zerons interacting to form a resonating entity. Matter consists of slow Zerons, held together by fast Zeron pressure. Inertia joins the strong nuclear force, gravitation and electro-magnetic force as a prime force of Nature. All have their origin in fast Zeron pressure. We have developed lucid models for many physical phenomena and discovered the meaning of Planck's Constant. By making time a non-variable Einstein's equations could be reworked resulting in the elimination of the counter-intuitive results of Relativity. As an extension of this we have developed a lucid model for mass increase with velocity. We have discovered a Universal Frame of Reference and found the underlying basis of Newton's laws. By using Zeron Theory the universe we live in can once more be explained in objectively real terms. More importantly we can now reduce all of this to a single simple proposition: The universe in all its complexity and diversity may be interpreted in terms of one particle, one force, and one law. Zeron theory has shown that physics is not based on the field concept or on other continuous structures. Physics is based on a discrete prime particle. We conclude that Einstein's castle in the air has collapsed.

General conclusions

It might be thought from the general tone of this book that the writer is somewhat anti-establishment. Nothing could be further from the truth. I just feel sorry for physicists who are mired in a system which makes no sense at all and which has frustrated them at every turn. In spite of the huge disability of not knowing about Zerons or the Cosma, the fact that scientists have been able to make such enormous strides is a monument to their brilliance and ingenuity. What might they have achieved *with* this knowledge!

The God Particle

What have we landed up with as a result of the Zeron theory? We have found the all pervading but totally reclusive "God Particle" in the form of the Zeron. We have found that matter is an agglomeration of these extremely small slow-Zerons held together by Zeron pressure caused by the bombardment of countless other fast-Zerons. Some of the fast-Zerons impact and are absorbed. An equal number get ejected. Each atom constantly exchanges the diminutive parts from which it is made. Matter only therefore exists as a sort of statistical entity, having no really permanent components.

All the forces of nature have their root in Zeron pressure. All forms of true repulsion are as a result of two resonating systems impinging on one another. There is no force of attraction. Attraction is always repulsion in the opposite direction. The strong nuclear force, gravitation, magnetism, and inertia can all be modeled on Zeron pressure. The last named is caused in accelerating systems as opposed to the other three which can be steady state systems. Inertia now joins the fundamental forces of nature in its own right. The weak nuclear force falls away.

We have lucid explanations for many of the basic natural phenomena, and have discovered among other things, what Planck's constant means, how Einstein did not include the term cos phi in his equations and the profound effect on his theories as a result of the inclusion of this term.

We have re-defined time as being invariable, developed an alternative real and objective version of relativity, have discovered that light can exceed the generally accepted value c, and have uncovered a Universal Frame of Reference.

Most importantly, we have discovered lucid reasons for debunking the Uncertainty Principle that has lead to such bizarre interpretations in conventional physics. The universe we live in once again becomes explainable in basic mechanistic terms. Even Newton's laws become more than empirical equations, and it is shown that there are fundamentals underlying these equations.

The many models constructed above, that inter-link in a satisfactory manner to present objectively real descriptions of the associated physical phenomena, constitute compelling evidence for the veracity of Zeron theory. In particular, Zeron theory provides a platform for the enunciation of a unified theory of matter and the basic forces of nature. Unified theory can be enunciated in terms of only three propositions:

- The Zeron is the fundamental constituent of all matter. (Particle)

- All forces in nature are a manifestation of Zeron pressure. (Force)

- The properties of the Zeron underlie all physical phenomena (Law)

The universe in all its complexity and diversity may be interpreted in terms of one particle, one force, and one law.

What now?

Are there practical implications for us as a result of Zeron theory? Can we now construct an anti-gravity machine using this knowledge, or perhaps construct a machine that would neutralize inertia? Imagine a motorcar that could accelerate to 100 km/hr in a fraction of a second and could make right angle turns at high speed! I'm afraid it's likely to be wishful thinking at least in the shorter term.

In spite of being blissfully ignorant about our hidden Zeron universe, scientists have very successfully pursued the quest for knowledge using the theoretical tools they had developed, however flawed. Whether the physicist perceives the world in terms of mathematical abstractions or quantum waves, or whatever, enough conventional knowledge has become available to interpret most physical phenomena adequately and to produce multitudinous practical applications from this data. If there were now to be an awareness of the existence of the Zeron world, I do not think that much would change in the short term. This hidden universe is, in the main, invisible and undetectable and it seems unlikely that there would be any immediate outworking simply because we know it is there. Perhaps this is a pessimistic view. Would it were so.

However, it is in the realm of understanding the *processes* behind our perceptions that Zeron theory creates a "quantum leap" forward. For instance, reach out and touch the top of a table. The hand stops as it meets the table surface. One can exert considerable force, yet the hand does not sink in. Why? At Zeron level millions of Zeron resonance shells on the surface of the hand collide with millions more at the surface of the table. These resonances repel each other. The harder we push the harder the force of repulsion. This repulsion is what we perceive as the resistance we feel with our hand. Our perception has a very real and describable foundation.

Using Zeron theory, detailed models such as these can be constructed for any physical phenomena, especially the more familiar ones that are part of our everyday lives. That is first prize!

Quo Vadis?

Where do we go from here? Clearly in covering such a vast field in so few pages has meant that we have only just touched the surface. If one accepts the basic premises of Zeron theory, it is clear that there is much to be done. Pick up any physics textbook and one will find huge amounts of theory that need to be re-interpreted in terms of Zeron physics.

Will this be done? The writer would certainly need many more lifetimes to accomplish even a small part of this development, and so it needs the effort of other better-qualified people to examine conventional physics and revise it. The hope is that during this process practical applications might be discovered. This should surely be the aim of all theoretical work. Free energy? Cold fusion? Anti-gravity? Who knows what a better understanding of the fundamentals of physics might bring?

There is also much basic work to be done on the Zeron theory itself. The areas that are less well developed in this document are those of "spin", molecular bonding and magnetism. Magnetism, in particular, seems to be ripe for a good mathematical analysis. All these could do with attention from specialists more comfortable in these areas of expertise than I.

The End in Sight

We end with a profound statement from Einstein, perhaps revealing his own unease about the theories he had authored:

"I consider it quite possible that physics might not be based on the field concept, i.e., on continuous structures. In that case, nothing remains of my entire castle in the air, gravitation theory included, [and of] the rest of modern physics."

Indeed, Zeron theory has shown that physics is not based on the field concept or on other continuous structures. Physics is based on non-continuous discrete prime particles.

The castle in the air has collapsed!

Printed in the United Kingdom
by Lightning Source UK Ltd.
123545UK00001B/109/A